MATH MATTERS
for Adults

Fractions

Author
Karen Lassiter
Austin Community College
Austin, Texas

Consultants
Connie Eichhorn
Omaha Public Schools
Omaha, Nebraska

M. Gail Joiner Ward
Swainsboro Technical Institute
Swainsboro, Georgia

About the Author

Dr. Karen Lassiter is currently a mathematics instructor at Austin Community College. She is a former Senior Math Editor for Steck-Vaughn Company, and has done extensive work with standardized test preparation. Dr. Lassiter holds a Ph.D. in Educational Research, Testing, and Instructional Design and a bachelor's degree in mathematics and science education from Florida State University.

About the Consultants

Connie Eichhorn is a supervisor of adult education programs in the Omaha Public Schools. A former mathematics teacher and ABE/GED instructor, she earned an undergraduate degree in math at Iowa State University and is completing a doctoral program in adult education at the University of Nebraska. She conducts workshops on math instruction for teachers of adult basic education.

M. Gail Joiner Ward is presently working at Swainsboro Technical Institute as an adult education instructor. Previously she worked with adult learners in a multimedia lab through the School of Education at Georgia Southern University. She has bachelor's and master's degrees in art and early childhood education from Valdosta State College and Georgia Southern University.

Staff Credits

Executive Editor: Ellen Lehrburger
Editor: Margie Weaver
Design Manager: Pamela Heaney
Illustration Credits: Kristian Gallagher, David Griffin, Alan Klemp, Mike Krone
Photo Credits: p. 9 ©David Young-Wolff/PhotoEdit; p. 39 ©Tony Freeman/PhotoEdit; p. 73 ©Park Street; p. 107 ©Park Street; p. 127 ©Richard Hutchings/Photo Researchers; p. 145 ©Park Street.
Cover Design: Pamela Heaney

ISBN 0-8114-3651-9

Copyright © 1993 Steck-Vaughn Company

All rights reserved. No part of the material protected by this copyright may be reproduced or utilized in any form or by any means, electronic or mechanical, including photocopying, recording, or by any information storage and retrieval system, without permission in writing from the copyright owner. Requests for permission to make copies of any part of the work should be mailed to: Copyright Permissions, Steck-Vaughn Company, P.O. Box 26015, Austin, Texas 78755.

Printed in the United States of America.
1 2 3 4 5 6 7 8 9 CK 98 97 96 95 94 93

Contents

To the Learner 5
Fraction Skills Inventory 6-8

Unit 1

THE MEANING OF FRACTIONS

Getting Ready 9-10
Writing Fractions 11-13
Real-Life Application   At Home 14
Reducing Fractions 15
Reducing Fractions to Lowest Terms 16
Problem Solving: Using a Circle Graph ... 17-18
Raising Fractions to Higher Terms 19-20
Real-Life Application   On the Job 21
Mixed Review 22
Comparing Fractions 23-24
Finding a Common Denominator 25
Whole Numbers, Mixed Numbers,
  and Fractions 26-27
Comparing Whole Numbers, Mixed
  Numbers, and Fractions 28
Problem Solving: Using a Ruler 29-30
Improper Fractions 31
Recognizing Proper Fractions, Improper
  Fractions, Whole Numbers, and
  Mixed Numbers 32
Changing Improper Fractions to Whole
  or Mixed Numbers 33-34
Changing Mixed Numbers to Improper
  Fractions 35-36
Unit Review 37-38

Unit 2

ADDING FRACTIONS

Getting Ready 39-40
Adding Fractions with the Same
  Denominator 41-42
Real-Life Application   On the Job 43
Adding and Reducing Fractions 44

Adding Fractions and Changing
  to Whole Numbers 45
Adding Fractions and Changing
  to Mixed Numbers 46
Mixed Review 47
Real-Life Application   Time Off 48
Adding Mixed Numbers 49-50
Adding Mixed Numbers, Whole
  Numbers, and Fractions 51
Mixed Review 52
Problem Solving: Using a Ruler 53-54
Finding Common Denominators 55
Adding Fractions with Different
  Denominators 56-59
Mixed Review 60
Real-Life Application   Daily Living 61
Comparing Denominators 62
Adding Fractions with Different
  Denominators 63-64
Finding the Lowest Common
  Denominator 65
Mixed Review 66
Adding Mixed Numbers 67
Adding Mixed Numbers, Whole
  Numbers, and Fractions 68
Problem Solving: Using Time Records ... 69-70
Unit Review 71-72

Unit 3

SUBTRACTING FRACTIONS

Getting Ready 73-74
Subtracting Fractions with the Same
  Denominator 75-76
Subtracting and Reducing Fractions 77
Subtracting Mixed Numbers 78
Real-Life Application   On the Job 79
Subtracting Fractions from Mixed
  Numbers 80
Subtracting Whole Numbers from
  Mixed Numbers 81
Mixed Review 82
Problem Solving: Using a Ruler 83-84

Changing Whole Numbers to Mixed Numbers	85
Subtracting Fractions from One	86
Subtracting Fractions from Whole Numbers	87
Subtracting Fractions from Mixed Numbers	88-89
Subtracting Mixed Numbers from Mixed Numbers	90
Real-Life Application Time Off	91
Mixed Review	92
Subtracting Fractions with Different Denominators	93-95
Subtracting Fractions Using the LCD	96
Mixed Review	97
Problem Solving: Using a Map	98-99
Subtracting Mixed Numbers	100
Subtracting Mixed Numbers with Borrowing	101-102
Subtracting Mixed Numbers and Fractions	103
Unit Review	104-106

Unit 4

MULTIPLYING FRACTIONS

Getting Ready	107-108
Multiplying Fractions by Fractions	109-110
Real-Life Application On the Job	111
Multiplying Three Fractions	112
Multiplying Fractions Using Cancellation	113-114
Multiplying Three Fractions Using Cancellation	115
Mixed Review	116
Multiplying Whole Numbers and Fractions	117-118
Multiplying Mixed Numbers and Whole Numbers	119
Problem Solving: Figuring Overtime	120-121
Multiplying Mixed Numbers and Fractions	122
Multiplying Mixed Numbers and Mixed Numbers	123
Real-Life Application Time Off	124
Unit Review	125-126

Unit 5

DIVIDING FRACTIONS

Getting Ready	127-128
Dividing Fractions by Fractions	129-130
Dividing Whole Numbers by Fractions	131-132
Dividing Mixed Numbers by Fractions	133-134
Real-Life Application At the Store	135
Mixed Review	136
Dividing Fractions by Whole Numbers	137-138
Dividing Mixed Numbers by Whole Numbers	139
Dividing Mixed Numbers by Mixed Numbers	140
Problem Solving: Using a Sketch	141-142
Unit Review	143-144

Unit 6

PUTTING YOUR SKILLS TO WORK

Getting Ready	145-146
Choose an Operation: Changing Units of Measurement	147
Choose an Operation: Being a Consumer	148
Choose an Operation: Using a Line Graph	149
Multi-Step Problems: Using Measurement	150-151
Multi-Step Problems: Being a Consumer	152-153
Multi-Step Problems: Using a Time Record	154-155

Fractions Skills Inventory	156-158
Glossary	159-162
Answers and Explanations	163-207
What's Next?	208

TO THE LEARNER

The four books in the Steck-Vaughn series *Math Matters for Adults* are *Whole Numbers; Fractions; Decimals and Percents;* and *Measurement, Geometry, and Algebra.* They are written to help you understand and practice arithmetic skills, real-life applications, and problem-solving techniques.

This book contains features which will make it easier for you to work with fractions and to apply them to your daily life.

A Skills Inventory test appears at the beginning and end of the book.
- The first test shows you how much you already know.
- The final test can show you how much you have learned.

Each unit has several Mixed Reviews and a Unit Review.
- The Mixed Reviews give you a chance to practice the skills you have learned.
- The Unit Review helps you decide if you have mastered those skills.

There is also a glossary at the end of the book.
- Turn to the glossary to find the meanings of words that are new to you.
- Use the definitions and examples to help strengthen your understanding of terms used in mathematics.

The book contains answers and explanations for the problems.
- The answers let you check your work.
- The explanations take you through the steps used to solve the problems.

Fractions Skills Inventory

Reduce to lowest terms. If the fraction is in lowest terms, write LT.

1. $\dfrac{4}{8} =$
2. $\dfrac{9}{12} =$
3. $\dfrac{4}{15} =$
4. $\dfrac{10}{30} =$
5. $\dfrac{32}{48} =$

Raise to higher terms with the given denominator.

6. $\dfrac{1}{3} = \dfrac{}{9}$
7. $\dfrac{2}{5} = \dfrac{}{10}$
8. $\dfrac{3}{7} = \dfrac{}{21}$
9. $\dfrac{7}{10} = \dfrac{}{30}$
10. $\dfrac{4}{9} = \dfrac{}{36}$

Change to a whole or mixed number.

11. $\dfrac{4}{3} =$
12. $\dfrac{8}{2} =$
13. $\dfrac{12}{5} =$
14. $\dfrac{3}{1} =$
15. $\dfrac{18}{10} =$

Change to an improper fraction.

16. $5\dfrac{3}{4} =$
17. $2\dfrac{1}{2} =$
18. $4\dfrac{2}{3} =$
19. $7\dfrac{4}{5} =$
20. $9\dfrac{7}{10} =$

Add. Reduce if possible.

21. $\dfrac{1}{5} + \dfrac{2}{5} =$
22. $\dfrac{3}{8} + \dfrac{3}{8} =$
23. $\dfrac{9}{10} + \dfrac{7}{10} =$
24. $\dfrac{3}{7} + \dfrac{2}{7} + \dfrac{2}{7} =$

25. $4\dfrac{2}{9}$
    $+3\dfrac{3}{9}$

26. $2\dfrac{5}{12}$
    $+7\dfrac{1}{12}$

27. $1\dfrac{5}{6}$
    $+6\dfrac{5}{6}$

28. $\dfrac{3}{10}$
    $+9\dfrac{7}{10}$

29. $10\dfrac{11}{15}$
    $+\ \ 5$

30. $\dfrac{1}{2}$
    $+\dfrac{3}{4}$

31. $\dfrac{2}{3}$
    $+\dfrac{4}{7}$

32. $\dfrac{5}{9}$
    $+\dfrac{5}{27}$

33. $\dfrac{3}{8}$
    $+\dfrac{1}{6}$

34. $\dfrac{4}{5}$
    $+\dfrac{3}{4}$

Add. Reduce if possible.

35. $2\frac{3}{7}$
    $+1\frac{1}{6}$

36. 5
    $+8\frac{1}{2}$

37. $9\frac{2}{3}$
    $+6\frac{7}{15}$

38. $10\frac{9}{11}$
    $+\frac{4}{22}$

39. $4\frac{9}{10}$
    $+2\frac{5}{6}$

Subtract. Reduce if possible.

40. $\frac{4}{5} - \frac{2}{5} =$

41. $\frac{8}{9} - \frac{2}{9} =$

42. $\frac{2}{3} - \frac{2}{3} =$

43. $\frac{7}{8} - \frac{3}{8} =$

44. $10\frac{7}{11}$
    $-3\frac{5}{11}$

45. $8\frac{9}{10}$
    $-\frac{7}{10}$

46. $6\frac{5}{6}$
    $-5\frac{1}{6}$

47. $9\frac{11}{12}$
    $-\frac{3}{12}$

48. $15\frac{17}{25}$
    -8

49. 1
    $-\frac{3}{5}$

50. 4
    $-\frac{7}{8}$

51. $7\frac{1}{3}$
    $-\frac{2}{3}$

52. $9\frac{2}{7}$
    $-8\frac{5}{7}$

53. $1\frac{4}{9}$
    $-\frac{8}{9}$

54. $\frac{3}{4}$
    $-\frac{1}{2}$

55. $\frac{7}{10}$
    $-\frac{3}{5}$

56. $\frac{7}{8}$
    $-\frac{7}{12}$

57. $\frac{1}{3}$
    $-\frac{1}{4}$

58. $\frac{5}{6}$
    $-\frac{4}{7}$

59. $8\frac{2}{3}$
    $-6\frac{1}{6}$

60. $5\frac{1}{4}$
    $-\frac{3}{5}$

61. $3\frac{5}{12}$
    $-2\frac{7}{10}$

62. $1\frac{2}{9}$
    $-\frac{11}{18}$

Multiply. Reduce if possible.

63. $\frac{1}{2} \times \frac{2}{3} =$

64. $\frac{3}{8} \times \frac{4}{9} =$

65. $\frac{5}{6} \times \frac{3}{4} =$

66. $\frac{2}{3} \times \frac{1}{4} \times \frac{3}{5} =$

67. $\frac{2}{9} \times 3 =$

68. $5 \times \frac{3}{10} =$

69. $8\frac{1}{2} \times 4 =$

70. $7 \times 3\frac{2}{3} =$

71. $1\frac{5}{9} \times \frac{1}{2} =$

72. $\frac{2}{3} \times 2\frac{2}{5} =$

73. $5\frac{1}{4} \times 1\frac{1}{7} =$

74. $4\frac{1}{6} \times 5\frac{4}{10} =$

Divide. Reduce if possible.

75. $\frac{4}{5} \div \frac{3}{10} =$

76. $\frac{5}{6} \div \frac{1}{6} =$

77. $\frac{2}{3} \div \frac{3}{5} =$

78. $\frac{7}{8} \div \frac{7}{16} =$

79. $9 \div \frac{1}{3} =$

80. $4 \div \frac{1}{4} =$

81. $10\frac{1}{2} \div \frac{2}{6} =$

82. $5\frac{1}{7} \div \frac{3}{7} =$

83. $\frac{1}{2} \div 3 =$

84. $\frac{2}{3} \div 7 =$

85. $2\frac{3}{8} \div 4 =$

86. $9\frac{3}{10} \div 3 =$

87. $12\frac{3}{5} \div 7 =$

88. $3\frac{3}{8} \div 1\frac{1}{8} =$

89. $15\frac{5}{9} \div 5\frac{1}{3} =$

90. $21\frac{2}{3} \div 3\frac{1}{2} =$

Below is a list of the problems in this Skills Inventory and the pages on which the skills are taught. If you missed any problems, turn to the pages listed and practice the skills. Then correct the problems you missed in the Skills Inventory.

Problem	Practice Page
Unit 1	
1-5	15-16
6-10	19-20
11-15	33-34
16-20	35-36
Unit 2	
21-24	41-42, 44-46
25-29	49-51
30-34	56-59, 63-65
35-39	67-68

Problem	Practice Page
Unit 3	
40-43	75-77
44-48	78, 80-81
49-53	86-90
54-58	93-96
59-62	100-103

Problem	Practice Page
Unit 4	
63-66	109-110, 112-115
67-70	117-119
71-74	122-123
Unit 5	
75-78	129-130
79-82	131-134
83-87	137-139
88-90	140

Unit 1 — The Meaning of Fractions

We use fractions to name a part of something. You may need them to find the total hours you worked, to follow a recipe, or to fix your car.

In this unit, you will learn how to recognize, write, and compare fractions and mixed numbers.

Getting Ready

You should be familiar with the skills on this page and the next before you begin this unit. To check your answers, turn to page 163.

 To add, line up the digits. Then add, beginning with the ones column.

Add.

1. 23 + 47 =

   23
   + 47

   70

2. 19 + 32 =

3. 153 + 9 =

4. 78 + 65 =

5. 314 + 611 =

6. 236 + 159 =

7. 97 + 423 =

8. 808 + 32 =

For review, see pages 41–43 in Math Matters for Adults, Whole Numbers.

Getting Ready

 To subtract a larger number from a smaller number in the same column, rename by borrowing from the column to the left.

Subtract.

9.
35 − 8 =

$$\begin{array}{r} 35 \\ -\ 8 \\ \hline 27 \end{array}$$

10.
41 − 16 =

11.
63 − 28 =

12.
80 − 12 =

For review, see pages 66–69 in **Math Matters for Adults, Whole Numbers.**

 To multiply a two-digit number by a one-digit number, first multiply by the ones digit. Carry if necessary. Then multiply the tens.

Multiply.

13.
22 × 3 =

$$\begin{array}{r} 22 \\ \times\ 3 \\ \hline 66 \end{array}$$

14.
46 × 6 =

15.
61 × 2 =

16.
83 × 5 =

For review, see pages 89–90 in **Math Matters for Adults, Whole Numbers.**

 To divide, begin with the first place of the number you're dividing into. Then work to the right. Use these steps: divide, multiply, subtract, and bring down. If there's a remainder, write R and the amount left over.

Divide.

17.
53 ÷ 4 =

$$\begin{array}{r} 13\ R1 \\ 4\overline{)53} \\ -\ 4 \\ \hline 13 \\ -12 \\ \hline 1 \end{array}$$

18.
38 ÷ 2 =

19.
29 ÷ 5 =

20.
42 ÷ 6 =

For review, see pages 127–131 in **Math Matters for Adults, Whole Numbers.**

Writing Fractions

A fraction names part of a whole or of a group. There are ten pins on a bowling alley. One bowling pin is 1 of 10, or $\frac{1}{10}$ of a group of ten bowling pins. There are twelve months in a year. One month is 1 of 12 or $\frac{1}{12}$ of a year.

Three of the four parts of the square are shaded.

fraction bar → $\frac{3}{4}$ ← numerator
← denominator

Use These Steps

Write a fraction for the parts that are shaded.

1. Make a fraction bar. Count the number of equal parts in the whole. This number is the denominator. Write the denominator under the fraction bar.

   $\frac{}{8}$

2. Count the number of shaded parts. This number is the numerator. Write the numerator over the fraction bar.

   $\frac{3}{8}$

Write the fraction for each shaded part.

1.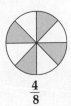
   $\frac{4}{8}$

2.

3.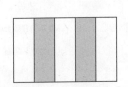

4.

5.

6.

7.

8.

Shade each figure to show the fraction.

9.
   $\frac{3}{4}$

10.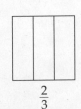
    $\frac{2}{3}$

11.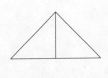
    $\frac{1}{2}$

12.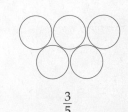
    $\frac{3}{5}$

Writing Fractions

You can also write fractions as words.

The circle below is divided into five equal parts. $\frac{1}{5}$ or one fifth of the circle is shaded.

Use the words in the box to help you when you write fractions.

half	third	fourth
fifth	sixth	seventh
eighth	ninth	tenth

Use These Steps

Write in words the fraction for the shaded parts.

1. Count how many equal parts the square is divided into.

   four

2. Count how many parts of the square are shaded.

   two fourths

Write in words the fraction for the shaded parts.

1.

   one half

2.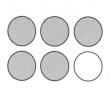

3.

4.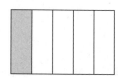

5.

6.

7.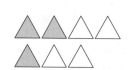

8.

9.

10.

11.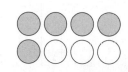

12.

Writing Fractions

We use fractions in everyday situations such as measuring with a ruler.

There are 12 inches in a foot. What fraction of a foot is 9 inches?

$$\frac{9}{12} \leftarrow \text{number of parts} \atop \leftarrow \text{parts in the whole}$$

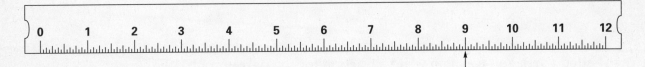

Use These Steps

There are 16 ounces in a pound. What fraction of a pound is 7 ounces?

1. Write the parts in the whole as the denominator.

    $\overline{16}$

2. Write the number of parts given as the numerator.

    $\frac{7}{16}$

Write a fraction for each of the parts described.

1. There are 12 months in a year. What fraction of a year is 3 months?

    Answer $\frac{3}{12}$

2. There are 36 inches in a yard. What fraction of a yard is 25 inches?

    Answer _____

3. There are 4 cups in a quart. What fraction of a quart is 1 cup?

    Answer _____

4. There are 3 feet in a yard. What fraction of a yard is 2 feet?

    Answer _____

5. There are 10 dimes in a dollar. What fraction of a dollar is 9 dimes?

    Answer _____

6. There are 60 minutes in an hour. What fraction of an hour is 42 minutes?

    Answer _____

7. There are 24 hours in a day. What fraction of a day is 17 hours?

    Answer _____

8. There are 12 eggs in a dozen. What fraction of a dozen is 12 eggs?

    Answer _____

Real-Life Application — At Home

Most cakes are in the shape of a circle or a rectangle. The pictures below show four round cakes and one rectangular cake.

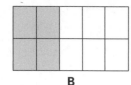

A   B   C   D   E

You can write a fraction for the part of each cake that is still in the pan (the shaded part). You can also write a fraction for the part of each cake that was eaten (the part that isn't shaded). Together these two fractions make up the whole cake.

Example   Find the cake that was cut into 10 pieces, or tenths. Write the fraction that shows how much of the cake is still in the pan.

Cake B is cut into 10 equal parts (tenths). 6 of the 10 parts have been eaten. 4 of the 10 parts are still in the pan.

$\frac{4}{10}$ of the cake is still in the pan.

Answer the following questions about the cakes.

1. Find the cake that was cut into fifths. Write the fraction that shows how much cake is left in the pan (the shaded part).

   Answer _____

2. Find the cake that was cut into eighths. Write the fraction that shows how much of the cake has been eaten (the part that isn't shaded).

   Answer _____

3. Write the fraction that shows how much cake is left of the one that was cut into sixths.

   Answer _____

4. Write the fraction that shows how much cake has been eaten of the one that was cut into fourths.

   Answer _____

5. Write the fraction that shows how much cake has been eaten of the one that was cut into tenths.

   Answer _____

6. Write the fraction that shows how much is left of the cake that was cut into eighths.

   Answer _____

Reducing Fractions

Reducing a fraction to lowest terms means dividing both the numerator and denominator by the same number. Sometimes you may have to reduce more than once to get the fraction in lowest terms. A fraction is reduced to lowest terms when you can divide both the numerator and denominator only by 1.

$$\frac{6}{18} = \frac{6 \div 2}{18 \div 2} = \frac{3 \div 3}{9 \div 3} = \frac{1}{3} \quad \text{or} \quad \frac{6}{18} = \frac{6 \div 6}{18 \div 6} = \frac{1}{3}$$

Only 1 divides into 1 and 3 evenly. $\frac{1}{3}$ is in lowest terms.

Use These Steps

Reduce $\frac{14}{28}$ to lowest terms.

1. Find a number that divides into 14 and 28 evenly.

   $$\frac{14}{28} = \frac{14 \div 2}{28 \div 2} = \frac{7}{14}$$

2. If any number other than 1 goes into 7 and 14 evenly, the number is not in lowest terms. Reduce again.

   $$\frac{7}{14} = \frac{7 \div 7}{14 \div 7} = \frac{1}{2}$$

3. See if you can divide both the numerator and the denominator by a number other than 1.

   $\frac{1}{2}$ is in lowest terms

Reduce each fraction to lowest terms.

1. $\frac{8}{12} =$

   $\frac{8 \div 2}{12 \div 2} = \frac{4}{6} = \frac{4 \div 2}{6 \div 2} = \frac{2}{3}$

2. $\frac{9}{21} =$

3. $\frac{14}{20} =$

4. $\frac{18}{24} =$

5. $\frac{25}{40} =$

6. $\frac{24}{30} =$

7. $\frac{36}{48} =$

8. $\frac{20}{24} =$

9. $\frac{12}{16} =$

10. $\frac{36}{40} =$

11. $\frac{8}{10} =$

12. $\frac{12}{18} =$

13. $\frac{7}{21} =$

14. $\frac{10}{12} =$

15. $\frac{8}{16} =$

16. $\frac{15}{25} =$

17. $\frac{24}{48} =$

18. $\frac{13}{39} =$

Reducing Fractions to Lowest Terms

In some fractions, both the numerator and denominator end in zeros. When this happens, you can use a short cut to reduce the fractions. Cross out an equal number of zeros on the top and on the bottom of the fraction.

The fraction that is left may need to be reduced again. When 1 is the only number you can divide the numerator and denominator by, the fraction is reduced to lowest terms.

Use These Steps

Reduce $\frac{20}{400}$ to lowest terms.

1. Cross out the zero on the top and 1 zero on the bottom.

    $\frac{2\cancel{0}}{40\cancel{0}}$

2. Reduce the fraction that is left. Find a number that will divide into both 2 and 40 evenly. Try 2.

    $\frac{2}{40} = \frac{2 \div 2}{40 \div 2} = \frac{1}{20}$

3. See if you can divide both the numerator and the denominator by a number other than 1.

    $\frac{1}{20}$ is in lowest terms

Reduce each fraction to lowest terms. If the fraction is already in lowest terms, write LT.

1. $\frac{6\cancel{0}}{10\cancel{0}} =$
   $\frac{6}{10} = \frac{6 \div 2}{10 \div 2} = \frac{3}{5}$

2. $\frac{9}{25} =$

3. $\frac{4}{6} =$

4. $\frac{30}{120} =$

5. $\frac{30}{70} =$

6. $\frac{10}{100} =$

7. $\frac{17}{28} =$

8. $\frac{300}{730} =$

9. $\frac{7}{30} =$

10. $\frac{8}{24} =$

11. $\frac{100}{150} =$

12. $\frac{5}{11} =$

13. $\frac{1}{9} =$

14. $\frac{200}{1400} =$

15. $\frac{6}{13} =$

16. $\frac{30}{50} =$

17. Jack's gas tank holds 20 gallons. He put 10 gallons into the tank. What fraction shows how much gas he put into the gas tank?

    Answer _____

18. John needs $\frac{6}{8}$ cup carrots to make a carrot cake. What is this amount in lowest terms?

    Answer _____

Problem Solving: Using a Circle Graph

A circle graph shows a total amount divided into parts. Each part is a fraction of the total.

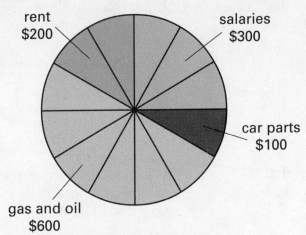

Expenses for Diaz Garage

The circle graph above shows the expenses paid by Manuel Diaz to operate his garage. He has $1,200 in total expenses.

Example What fraction of Manuel's expenses is for rent? Reduce the fraction to lowest terms.

▶ **Step 1.** Write the total expenses, $1,200, as the denominator of a fraction.

$$\frac{}{1200}$$

▶ **Step 2.** Write the amount paid for rent, $200, as the numerator of the fraction.

$$\frac{200}{1200}$$

▶ **Step 3.** Reduce. Cross out an equal number of zeros on the end of the numerator and the denominator. Reduce again.

$$\frac{2\cancel{00}}{12\cancel{00}} = \frac{2}{12} = \frac{1}{6}$$

$\frac{1}{6}$ of Manuel's expenses is for rent.

Solve. Reduce to lowest terms.

1. Write the fraction that shows what part of Manuel's expenses is for employees' salaries.

   Answer _____

2. What fraction shows the part that Manuel spends on gasoline and oil?

   Answer _____

Solve. Reduce to lowest terms.

3. What fraction shows the part that Manuel spends on car parts?

   Answer_____

4. What fraction shows the part that Manuel spends on parts and gas and oil together?
   (Hint: Add the two amounts first. Then write a fraction.)

   Answer_____

5. What fraction shows the part that Manuel spends on rent and salaries together?

   Answer_____

6. What fraction shows the total that Manuel spends on all of his expenses? (Hint: Add all the amounts. Then write a fraction.)

   Answer_____

7. Manuel spends a total of $300 on salaries. He pays his mechanic $150. Write the fraction that shows what part of the total salary expense he pays for his mechanic's salary.

   Answer_____

8. Next year Manuel's rent will go up to $250. His total expenses will be $1,250. Write the fraction that shows what part of Manuel's expenses he will have to pay for rent.

   Answer_____

9. Complete the table below using the circle graph on page 17.

Expense	Amount	Fraction
rent	$200	$\frac{200}{1200} = \frac{1}{6}$
salaries		
car parts		
gas and oil		

Raising Fractions to Higher Terms

When you change a fraction to an equal fraction with a larger denominator, it is called raising a fraction to higher terms.

To raise a fraction to higher terms, multiply both the numerator and the denominator by the same number.

$$\frac{3}{5} = \frac{3 \times 3}{5 \times 3} = \frac{9}{15}$$

Use These Steps

Raise $\frac{2}{5}$ to higher terms by multiplying the numerator and denominator by 6.

1. Multiply the numerator by 6.

2. Multiply the denominator by 6.

Write equal fractions for each figure.

1.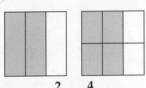

   $\frac{2}{3} = \frac{4}{6}$

2.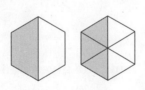

3.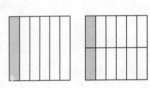

4.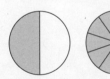

5.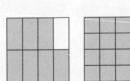

6.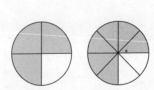

Multiply the numerator and denominator by 4 to raise each fraction to higher terms.

7. $\frac{2}{5} = \frac{2 \times 4}{5 \times 4} = \frac{8}{20}$

8. $\frac{3}{8} =$

9. $\frac{1}{6} =$

10. $\frac{4}{5} =$

Multiply the numerator and denominator by 5 to raise each fraction to higher terms.

11. $\frac{3}{4} = \frac{3 \times 5}{4 \times 5} = \frac{15}{20}$

12. $\frac{4}{7} =$

13. $\frac{2}{3} =$

14. $\frac{1}{9} =$

19

Raising Fractions to Higher Terms

You can raise a fraction to higher terms to get a new fraction with a given denominator.

Use These Steps

Raise $\frac{1}{8}$ to higher terms with a denominator of 32.

1. Write the fraction, $\frac{1}{8}$. Write another fraction with the new denominator, 32.

   $$\frac{1}{8} = \frac{}{32}$$

2. To find the number to multiply by, think about the old denominator, 8, and the new denominator, 32. 8 multiplied by what number equals 32?

   $$\frac{1}{8} = \frac{}{8 \times 4} = \frac{}{32}$$

3. Multiply the numerator, 1, by 4.

   $$\frac{1}{8} = \frac{1 \times 4}{8 \times 4} = \frac{4}{32}$$

Write the number used to raise each fraction to higher terms.

1. $\frac{1}{2} = \frac{3}{6}$ __3__

2. $\frac{3}{4} = \frac{12}{16}$ ____

3. $\frac{7}{8} = \frac{21}{24}$ ____

4. $\frac{5}{9} = \frac{30}{54}$ ____

5. $\frac{6}{10} = \frac{18}{30}$ ____

6. $\frac{12}{15} = \frac{48}{60}$ ____

7. $\frac{2}{9} = \frac{10}{45}$ ____

8. $\frac{5}{12} = \frac{25}{60}$ ____

Raise each fraction to higher terms using the given denominator.

9. $\frac{3}{7} = \frac{\boxed{9}}{21}$

10. $\frac{2}{14} = \frac{\boxed{}}{56}$

11. $\frac{15}{25} = \frac{\boxed{}}{100}$

12. $\frac{9}{16} = \frac{\boxed{}}{48}$

13. $\frac{18}{45} = \frac{\boxed{}}{90}$

14. $\frac{24}{30} = \frac{\boxed{}}{90}$

15. $\frac{1}{9} = \frac{\boxed{}}{54}$

16. $\frac{7}{11} = \frac{\boxed{}}{77}$

17. $\frac{14}{28} = \frac{\boxed{}}{84}$

18. $\frac{5}{24} = \frac{\boxed{}}{96}$

19. $\frac{3}{21} = \frac{\boxed{}}{63}$

20. $\frac{8}{12} = \frac{\boxed{}}{72}$

21. Kevin read 10 pages of a 25-page book. Beverly read an equal fraction of a 50-page book. How many pages did Beverly read?

    Answer_____

22. Carmen typed 14 pages of a 30-page term paper. Scott typed an equal fraction of a 60-page paper. How many pages did Scott type?

    Answer_____

Real-Life Application — On the Job

Ken works in a candy factory. He weighs ingredients on a large scale. He writes weights as fractions.

Use the information in the box to help answer the questions below.

> 16 ounces = 1 pound
> 2,000 pounds = 1 ton

Example Ken weighed three drums of sugar. Together they weighed 1,500 pounds. What fraction of a ton is this?

$$\frac{15\cancel{00}}{20\cancel{00}} = \frac{15}{20} = \frac{3}{4}$$

The drums weigh $\frac{3}{4}$ ton.

Solve. Reduce to lowest terms.

1. Abram put six bags of peanuts on the scale. All together they weighed 1,000 pounds. What fraction of a ton was on the scale?

    Answer_____

2. José weighs the candy after it is made. He packed ten cartons of peanut brittle and put them on the scale. The cartons weighed 700 pounds. What fraction of a ton did the ten cartons of peanut brittle weigh?

    Answer_____

3. Ellen packs boxes of caramels. She has a small scale that measures pounds. Each box of caramels weighs 14 ounces. What fraction of a pound is this?

    Answer_____

4. Nora packs boxes of chocolates. Each box weighs 8 ounces. What fraction of a pound is each box?

    Answer_____

5. Keiko prepares the ingredients for each batch of fudge. Each batch requires 100 pounds of sugar. What fraction of a ton is 100 pounds?

    Answer_____

6. Keiko packs the fudge in 12-ounce boxes. What fraction of a pound is each box of fudge?

    Answer_____

Mixed Review

Write the fraction and the word name for the shaded part of each figure.

1.

2.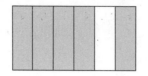

3.

4.

_____ _____ _____ _____

_____ _____ _____ _____

5. There are 12 inches in a foot. What fraction of a foot is 5 inches?

   Answer _____

6. There are 7 days in a week. What fraction of a week is 2 days?

   Answer _____

7. There are 24 quarts of oil in a full case. What fraction of a case is 13 quarts?

   Answer _____

8. There are 52 weeks in a year. What fraction of a year is 3 weeks?

   Answer _____

Reduce each fraction to lowest terms. If the fraction is in lowest terms, write LT.

9. $\frac{15}{25} =$

10. $\frac{21}{49} =$

11. $\frac{36}{54} =$

12. $\frac{81}{90} =$

13. $\frac{7}{10} =$

14. $\frac{10}{50} =$

15. $\frac{20}{40} =$

16. $\frac{70}{210} =$

17. $\frac{50}{100} =$

18. $\frac{200}{1600} =$

Raise each fraction to higher terms using the given denominator.

19. $\frac{1}{2} = \frac{}{4}$

20. $\frac{2}{3} = \frac{}{9}$

21. $\frac{4}{5} = \frac{}{15}$

22. $\frac{3}{7} = \frac{}{21}$

23. $\frac{5}{6} = \frac{}{30}$

24. $\frac{3}{4} = \frac{}{16}$

25. $\frac{5}{8} = \frac{}{24}$

26. $\frac{7}{10} = \frac{}{50}$

Comparing Fractions

You can use a number line to compare fractions. The fraction that is farther to the right is the greater fraction.

Remember the symbol > means *greater than*. $\frac{1}{2} > \frac{1}{4}$

　　　　　　　　< means *less than*. $\frac{5}{8} < \frac{7}{8}$

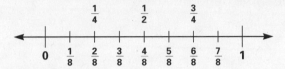

Use These Steps

Compare $\frac{5}{8}$ and $\frac{1}{2}$.

1. Find the fractions on the number line. Write the fractions.

2. Decide which fraction is farther to the right on the number line. Write the correct symbol between the fractions.

$\frac{5}{8}$　$\frac{1}{2}$　　　　　　$\frac{5}{8} > \frac{1}{2}$

Use the number lines to compare each set of fractions. Write >, <, or =.

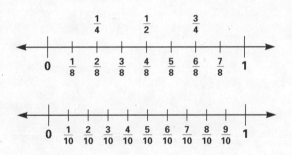

1. $\frac{3}{8}$ > $\frac{2}{10}$

2. $\frac{5}{8}$ < $\frac{7}{8}$

3. $\frac{5}{10}$ = $\frac{4}{8}$

4. $\frac{3}{8}$ ☐ $\frac{5}{8}$

5. $\frac{3}{4}$ ☐ $\frac{7}{10}$

6. $\frac{4}{10}$ ☐ $\frac{5}{8}$

7. $\frac{1}{8}$ ☐ $\frac{1}{10}$

8. $\frac{3}{8}$ ☐ $\frac{3}{10}$

9. $\frac{1}{2}$ ☐ $\frac{5}{10}$

10. $\frac{7}{10}$ ☐ $\frac{7}{8}$

11. $\frac{9}{10}$ ☐ $\frac{3}{4}$

12. $\frac{1}{4}$ ☐ $\frac{1}{10}$

Comparing Fractions

Fractions can be compared without using a number line. If the denominators are the same, then compare the numerators.

$$\frac{3}{8} < \frac{5}{8}$$

If the denominators are not the same, you need to first change the fractions to equal fractions with the same denominators.

Use These Steps

Compare $\frac{5}{8}$ and $\frac{3}{4}$.

1. The denominators are not the same. Change $\frac{3}{4}$ to a fraction with 8 as the denominator.

    $$\frac{3}{4} = \frac{3 \times 2}{4 \times 2} = \frac{6}{8}$$

2. Compare the numerator of each fraction. 5 is less than 6.

    $$\frac{5}{8} < \frac{6}{8}, \text{ so } \frac{5}{8} < \frac{3}{4}$$

Compare each pair of fractions. Write >, <, or =.

1. $\frac{2}{5} \boxed{=} \frac{6}{15}$
   $\frac{2}{5} = \frac{2 \times 3}{5 \times 3} = \frac{6}{15}$

2. $\frac{3}{8} \square \frac{1}{2}$

3. $\frac{2}{4} \square \frac{3}{8}$

4. $\frac{4}{5} \square \frac{2}{10}$

5. $\frac{1}{2} \square \frac{6}{10}$

6. $\frac{4}{9} \square \frac{2}{3}$

7. $\frac{1}{4} \square \frac{4}{12}$

8. $\frac{5}{6} \square \frac{2}{3}$

9. $\frac{1}{4} \square \frac{3}{8}$

10. $\frac{2}{7} \square \frac{5}{14}$

11. $\frac{3}{4} \square \frac{7}{16}$

12. $\frac{4}{7} \square \frac{10}{21}$

13. $\frac{5}{8} \square \frac{13}{16}$

14. $\frac{7}{12} \square \frac{1}{2}$

15. $\frac{2}{5} \square \frac{10}{15}$

16. $\frac{3}{4} \square \frac{16}{20}$

17. $\frac{7}{9} \square \frac{15}{18}$

18. $\frac{3}{10} \square \frac{10}{20}$

19. $\frac{5}{8} \square \frac{15}{24}$

20. $\frac{5}{10} \square \frac{1}{2}$

21. By lunch Pat completed $\frac{2}{3}$ of her work. Lena completed $\frac{5}{9}$ of her work. Who completed more of her work? (Hint: Which fraction is larger?)

22. Barb needs $\frac{1}{2}$ cup raisins to make cookies. She has $\frac{5}{8}$ cup raisins. Does she have enough for the cookies?

Answer _____

Answer _____

Finding a Common Denominator

To compare some fractions, you will need to change the denominators of both fractions. This is called finding a common denominator. A common denominator is a number that both denominators divide into evenly.

Use These Steps

Compare $\frac{1}{2}$ and $\frac{1}{3}$.

1. Multiply the two denominators to find a common denominator. $2 \times 3 = 6$. The new denominator for both fractions is 6.

    $\frac{1}{2} = \frac{}{2 \times 3} = \frac{}{6}$

    $\frac{1}{3} = \frac{}{3 \times 2} = \frac{}{6}$

2. Write each fraction in higher terms with 6 as the denominator.

    $\frac{1}{2} = \frac{1 \times 3}{2 \times 3} = \frac{3}{6}$

    $\frac{1}{3} = \frac{1 \times 2}{3 \times 2} = \frac{2}{6}$

3. Compare the numerators of the two new fractions. 3 is more than 2.

    $\frac{3}{6} > \frac{2}{6}$, so $\frac{1}{2} > \frac{1}{3}$

Compare each set of fractions. Write >, <, or =.

1. $\frac{1}{5}$ $\boxed{>}$ $\frac{1}{6}$

    $\frac{1}{5} = \frac{1 \times 6}{5 \times 6} = \frac{6}{30}$

    $\frac{1}{6} = \frac{1 \times 5}{6 \times 5} = \frac{5}{30}$

2. $\frac{2}{3}$ $\square$ $\frac{3}{4}$

3. $\frac{1}{2}$ $\square$ $\frac{3}{5}$

4. $\frac{5}{6}$ $\square$ $\frac{4}{7}$

5. $\frac{3}{4}$ $\square$ $\frac{3}{5}$

6. $\frac{4}{5}$ $\square$ $\frac{4}{7}$

7. $\frac{3}{8}$ $\square$ $\frac{2}{3}$

8. $\frac{7}{10}$ $\square$ $\frac{1}{3}$

9. $\frac{1}{2}$ $\square$ $\frac{5}{10}$

10. $\frac{3}{4}$ $\square$ $\frac{9}{10}$

11. $\frac{1}{3}$ $\square$ $\frac{1}{4}$

12. $\frac{2}{4}$ $\square$ $\frac{3}{6}$

13. Yesterday it rained $\frac{3}{10}$ inch in Orlando. It rained $\frac{3}{4}$ inch in Miami. Which city got more rain?

14. Peterboro received $\frac{3}{4}$ foot of snow on Sunday. Stowe received $\frac{1}{2}$ foot. Which city had more snow?

Answer _____

Answer _____

Whole Numbers, Mixed Numbers, and Fractions

Remember that a fraction shows a part of something. A whole number shows the whole of something. A mixed number shows both a whole and a part.

$\frac{1}{2}$

2

$2\frac{1}{2}$

Use These Steps

Write a mixed number for the shaded parts.

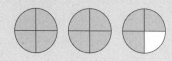

1. Count the number of whole circles that are shaded. Write the whole number.

    2

2. Count the number of parts of the remaining whole circle that are shaded.

    $2\frac{3}{4}$

Write a whole number or a mixed number for the shaded parts.

1.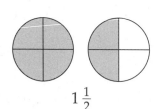

    $1\frac{1}{2}$

2.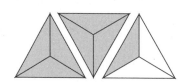

3.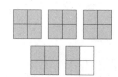

4.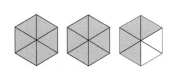

5.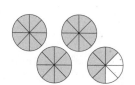

6.

7. Derrick used four full gallons of paint and one eighth of another gallon to paint his house. Write a mixed number to show the total amount of paint that he used.

    Answer _____

8. At the last family reunion, the Nealons ate three whole apple pies and one half a cherry pie. Write a mixed number to show how many pies the Nealons ate.

    Answer _____

Whole Numbers, Mixed Numbers, and Fractions

When you divide whole numbers, the answer may have a remainder. You can write the remainder as the numerator of a fraction. The number you divide by is the denominator. The answer can be written as a mixed number.

$$3\overline{)14} \quad 4 \text{ R2} = 4\frac{2}{3}$$
$$\underline{-12}$$
$$2$$

$$6\overline{)33} \quad 5 \text{ R3} = 5\frac{3}{6} = 5\frac{1}{2}$$
$$\underline{-30}$$
$$3$$

Use These Steps

Write the answer to $10 \div 4$ as a mixed number.

1. Divide.

$$4\overline{)10} \quad 2 \text{ R2}$$
$$\underline{-8}$$
$$2$$

2. Write the remainder as a fraction. Write the answer as a mixed number.

$$4\overline{)10} \quad 2 \text{ R2} = 2\frac{2}{4}$$
$$\underline{-8}$$
$$2$$

3. Reduce the fraction to lowest terms.

$$2\frac{2}{4} = 2\frac{1}{2}$$

Write each answer as a mixed number. Reduce the fraction if possible.

1. $15 \div 6 =$

$$6\overline{)15} \quad 2 \text{ R3} = 2\frac{3}{6} = 2\frac{1}{2}$$
$$\underline{-12}$$
$$3$$

2. $9 \div 4 =$

3. $10 \div 7 =$

4. $12 \div 5 =$

5. $20 \div 3 =$

6. $18 \div 8 =$

7. $17 \div 2 =$

8. $22 \div 10 =$

9. $36 \div 14 =$

10. $50 \div 20 =$

11. $72 \div 30 =$

12. $100 \div 15 =$

Comparing Whole Numbers, Mixed Numbers, and Fractions

To compare whole numbers, mixed numbers, and fractions, first compare the whole numbers. If the whole numbers are the same, then compare the fractions. Make sure that all the fractions have the same denominator.

$$6\frac{1}{2} > 2\frac{3}{4} \qquad 2\frac{3}{4} > 2\frac{1}{4} \qquad 2 < 2\frac{1}{2}$$

Use These Steps

Compare $1\frac{1}{2}$ and $1\frac{2}{5}$.

1. The whole numbers are the same. Find a common denominator for the fractions by multiplying the two denominators. $2 \times 5 = 10$. The new denominator for both fractions is 10.

   $1\frac{1}{2} = 1\frac{}{10}$
   $1\frac{2}{5} = 1\frac{}{10}$

2. Write each fraction in higher terms with 10 as the denominator for each fraction.

   $1\frac{1}{2} = 1\frac{5}{10}$
   $1\frac{2}{5} = 1\frac{4}{10}$

3. Compare the numerators of the two new fractions. 5 is greater than 4.

   $1\frac{5}{10} > 1\frac{4}{10}$,
   so $1\frac{1}{2} > 1\frac{2}{5}$

Compare. Write >, <, or =.

1. $2\frac{3}{8}$ $\boxed{>}$ $2\frac{1}{4}$
   $2\frac{3}{8} = 2\frac{3}{8}$
   $2\frac{1}{4} = 2\frac{2}{8}$

2. $3\frac{1}{3}$ $\square$ $4\frac{1}{8}$

3. $4\frac{1}{5}$ $\square$ $4\frac{2}{10}$

4. $5\frac{3}{4}$ $\square$ $5\frac{7}{8}$

5. $11\frac{1}{2}$ $\square$ $11\frac{1}{10}$

6. 7 $\square$ $7\frac{1}{5}$

7. $\frac{1}{5}$ $\square$ $1\frac{1}{5}$

8. $12\frac{3}{4}$ $\square$ $12\frac{15}{16}$

9. $2\frac{1}{2}$ $\square$ $2\frac{1}{12}$

10. $8\frac{7}{8}$ $\square$ $8\frac{9}{16}$

11. $5\frac{1}{3}$ $\square$ $5\frac{5}{15}$

12. $\frac{9}{10}$ $\square$ $10\frac{1}{10}$

Problem Solving: Using a Ruler

The ruler below is divided into inches and fractions of an inch: halves, fourths, eighths, and sixteenths.

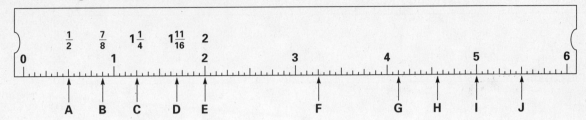

The arrow at A is $\frac{1}{2}$ inch from zero.

The arrow at B is $\frac{7}{8}$ inch from zero.

The arrow at C is $1\frac{1}{4}$ inches from zero.

The arrow at D is $1\frac{11}{16}$ inches from zero.

The arrow at E is 2 inches from zero.

Example How far from zero is the arrow at F?

▶ **Step 1.** Find the nearest whole inch mark to the left of the arrow at F.

$$3$$

▶ **Step 2.** Find the fraction of an inch the arrow is pointing to.

$$\frac{1}{4}$$

▶ **Step 3.** Write the mixed number.

$$3\frac{1}{4} \text{ inches}$$

Use the ruler above to answer each question.

1. How far from zero is the arrow at G?

   Answer_____

2. How far from zero is the arrow at H?

   Answer_____

3. How far from zero is the arrow at I?

   Answer_____

4. How far from zero is the arrow at J?

   Answer_____

Use the ruler below to answer each question.

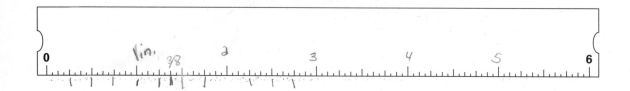

5. Find each inch mark on the ruler. Write the number of the inch above the inch mark.
   (Hint: Look at the 6-inch mark.)

6. Find 0 on the ruler. Write $\frac{1}{4}$ above the first $\frac{1}{4}$-inch mark to the right of 0.

7. Find the 1-inch mark on the ruler. Write $\frac{1}{2}$ above the first $\frac{1}{2}$-inch mark to the right of the 1-inch mark.

8. Find the 2-inch mark on the ruler. Write $\frac{1}{16}$ above the last $\frac{1}{16}$-inch mark before the 3-inch mark.

9. Find 1 inch on the ruler. Draw an arrow to the mark and label it A.

10. Find $1\frac{3}{8}$ inches on the ruler. Draw an arrow to the mark and label it B.

11. Find $2\frac{3}{16}$ inches on the ruler. Draw an arrow to the mark and label it C.

12. Find $3\frac{1}{4}$ inches on the ruler. Draw an arrow to the mark and label it D.

13. Find $4\frac{1}{2}$ inches on the ruler. Draw an arrow to the mark and label it E.

14. Find $4\frac{3}{4}$ inches on the ruler. Draw an arrow to the mark and label it F.

15. Find $5\frac{9}{16}$ inches on the ruler. Draw an arrow to the mark and label it G.

16. Find $5\frac{7}{8}$ inches on the ruler. Draw an arrow to the mark and label it H.

Reduce the following measurements to lowest terms. If the fraction is in lowest terms, write LT.

17. $3\frac{6}{16}$ inches =

18. $1\frac{2}{4}$ inches =

19. $\frac{6}{8}$ inch =

20. $4\frac{9}{16}$ inches =

21. $5\frac{4}{8}$ inches =

22. $2\frac{4}{16}$ inch =

Improper Fractions

A proper fraction is a fraction with a numerator that is smaller than the denominator.

An improper fraction is a fraction with a numerator equal to or larger than its denominator.

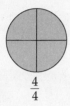
$\frac{4}{4}$

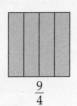
$\frac{9}{4}$

Use These Steps

Write an improper fraction for the number of pieces of pie remaining.

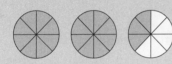

1. Count the number of equal parts in one pie. Write this number as the denominator.

   $\frac{}{8}$

2. Count the total number of remaining (shaded) pieces. Write this number as the numerator.

   $\frac{19}{8}$

Write an improper fraction for the shaded part of each group of figures.

1.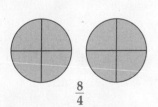
   $\frac{8}{4}$

2.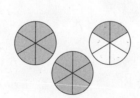

3.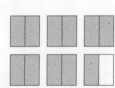

4.

5.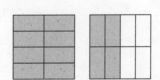

6.

7.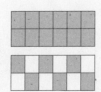

8.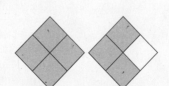

9.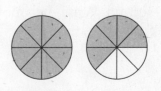

31

Recognizing Proper Fractions, Improper Fractions, Whole Numbers, and Mixed Numbers

A proper fraction has a numerator that is smaller than the denominator.

$$\frac{2}{3} \qquad \frac{3}{7}$$

An improper fraction has a numerator that is equal to or larger than the denominator.

$$\frac{8}{8} \qquad \frac{21}{6}$$

A whole number has no fraction part.

$$2 \qquad 15$$

A mixed number has a whole number and a fraction part.

$$3\frac{1}{2} \qquad 12\frac{2}{7}$$

Write P (proper fraction), I (improper fraction), W (whole number), or M (mixed number) on the line next to each problem.

1. $3\frac{1}{8}$ __M__
2. $\frac{2}{5}$ _____
3. $\frac{9}{1}$ _____
4. $\frac{10}{4}$ _____
5. $2\frac{1}{2}$ _____

6. 19 _____
7. $\frac{35}{6}$ _____
8. $\frac{3}{3}$ _____
9. 11 _____
10. $\frac{30}{30}$ _____

11. $\frac{5}{10}$ _____
12. $\frac{2}{45}$ _____
13. $8\frac{7}{8}$ _____
14. $\frac{61}{7}$ _____
15. 27 _____

16. $2\frac{4}{8}$ _____
17. $11\frac{1}{4}$ _____
18. $\frac{12}{3}$ _____
19. $\frac{13}{13}$ _____
20. $\frac{23}{4}$ _____

21. Walter bought two melons. He cut each melon into 8 slices. He ate all the slices of one melon and 3 slices of the second. Write the amount he ate as an improper fraction.

22. Write the amount that Walter ate as a mixed number.

Answer _____

Answer _____

Changing Improper Fractions to Whole or Mixed Numbers

You may need to change an improper fraction to a whole or mixed number. To change an improper fraction, divide the numerator by the denominator.

Use These Steps

Change $\frac{13}{6}$ to a mixed number.

1. Set up a division problem.

    $6\overline{)13}$

2. Divide.

    $\begin{array}{r} 2\text{ R1} \\ 6\overline{)13} \\ -12 \\ \hline 1 \end{array}$

3. Write the remainder as a fraction.

    $2\frac{1}{6}$

Write an improper fraction and a whole or mixed number for each group of figures.

1.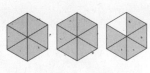

    $\frac{17}{6} = 2\frac{5}{6}$

2.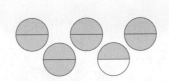

3.

4.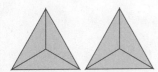

5.

6.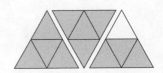

Change each improper fraction to a whole or mixed number.

7. $\frac{9}{4} = 2\frac{1}{4}$

    $\begin{array}{r} 2\text{ R1} \\ 4\overline{)9} \\ -8 \\ \hline 1 \end{array}$

8. $\frac{13}{9} =$

9. $\frac{4}{2} =$

10. $\frac{8}{5} =$

11. eighteen sixths = $\frac{18}{6} = 3$

    $\begin{array}{r} 3 \\ 6\overline{)18} \end{array}$

12. thirteen eighths =

13. twenty ninths =

14. sixteen sevenths =

Changing Improper Fractions to Whole or Mixed Numbers

Sometimes when you are changing an improper fraction to a mixed number, the remainder will be a fraction that needs to be reduced.

Use These Steps

Change $\frac{21}{6}$ to a mixed number.

1. Set up a division problem. Divide the numerator, 21, by the denominator, 6.

   $6\overline{)21}$

2. Divide. Write the remainder as a fraction.

   $6\overline{)21}$ R3 = $3\frac{3}{6}$
   -18
   3

3. Reduce the fraction by dividing the numerator and the denominator by the same number, 3.

   $3\frac{3}{6} = 3\frac{1}{2}$

Write each fraction as a whole or mixed number. Reduce fractions if possible.

1. $\frac{31}{6} =$
   $6\overline{)31}$ R1 = $5\frac{1}{6}$
   -30
   1

2. $\frac{6}{3} =$

3. $\frac{45}{9} =$

4. $\frac{67}{8} =$

5. $\frac{12}{5} =$

6. $\frac{34}{18} =$

7. $\frac{55}{9} =$

8. $\frac{28}{3} =$

9. $\frac{100}{45} =$

10. $\frac{18}{5} =$

11. $\frac{50}{10} =$

12. $\frac{38}{15} =$

13. $\frac{76}{30} =$

14. $\frac{42}{7} =$

15. $\frac{64}{9} =$

16. $\frac{81}{11} =$

17. $\frac{70}{3} =$

18. $\frac{32}{8} =$

19. $\frac{21}{4} =$

20. $\frac{8}{7} =$

21. Barry's Bakery cuts the pies it sells into 6 pieces each. The bakery sold 26 pieces of pie on Friday. How many pies did the bakery sell?

    Answer _____

22. The Oak Hill Youth Group used 42 eggs to make ice cream. If there are 12 eggs in a dozen, how many dozen eggs were used?

    Answer _____

Changing Mixed Numbers to Improper Fractions

Sometimes you may need to change a mixed number to an improper fraction.

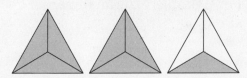

$2\frac{1}{3} = \frac{7}{3}$ ← number of shaded parts
 ← number of parts in the whole

Use These Steps

Write $3\frac{1}{4}$ as an improper fraction.

1. Write a fraction with the same denominator, 4.

   $3\frac{1}{4} = \frac{}{4}$

2. Multiply the denominator of the fraction, 4, by the whole number, 3.

   $4 \times 3 = 12$

3. Add the numerator of the fraction, 1, to 12. $12 + 1 = 13$. Write the sum over the denominator.

   $3\frac{1}{4} = \frac{13}{4}$

Write a mixed number for the shaded part. Change to an improper fraction.

1.

   $1\frac{1}{2} = \frac{3}{2}$

2.

3.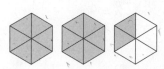

4.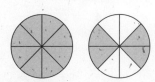

5.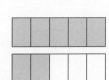

6.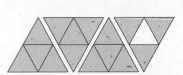

7.

8.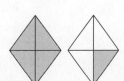

9.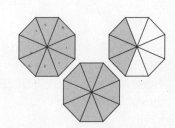

Changing Mixed Numbers to Improper Fractions

Remember that a fraction with the numerator equal to or larger than the denominator is an improper fraction. For example, $\frac{7}{5}$ is an improper fraction which is equal to $1\frac{2}{5}$.

To change a mixed number to an improper fraction, you need to keep the same denominator.

Use These Steps

Write $4\frac{1}{4}$ as an improper fraction with a denominator of 4.

1. Write the denominator, 4, for the improper fraction.

2. Multiply the denominator of the fraction, 4, by the whole number, 4.

3. Add the numerator of the fraction, 1, to 16. $16 + 1 = 17$. Write the sum over the denominator.

$$4\frac{1}{4} = \frac{}{4}$$

$$4 \times 4 = 16$$

$$4\frac{1}{4} = \frac{17}{4}$$

Write an improper fraction for each mixed number.

1. $3\frac{1}{3} = \frac{10}{3}$
   $3 \times 3 = 9$
   $9 + 1 = 10$

2. $6\frac{2}{5} = \frac{}{5}$

3. $13\frac{1}{8} = \frac{}{8}$

4. $12\frac{2}{8} = \frac{9}{8}$

5. $7\frac{1}{2} = 15$

6. $8\frac{2}{3} =$

7. $10\frac{7}{10} =$

8. $9\frac{5}{6} =$

9. $7\frac{1}{5} =$

10. $15\frac{6}{7} =$

11. $1\frac{1}{12} =$

12. $4\frac{1}{9} =$

13. Diane cut a 10-foot plank into 2-foot pieces. Write an improper fraction with 2 as the denominator.

    Answer _____

14. Mark made $4\frac{1}{2}$ quarts of punch for a party at work. Write an improper fraction to show how much punch Mark made.

    Answer _____

Unit 1 Review

Write a proper fraction for the shaded part of each figure.

1. $\frac{1}{3}$

2. $\frac{3}{5}$

3.  $\frac{5}{6}$

Write an improper fraction for the shaded part of each group of figures. Change each improper fraction to a whole or mixed number. Reduce all fractions to lowest terms.

4. 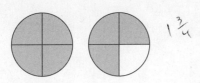 $1\frac{3}{4}$

5. 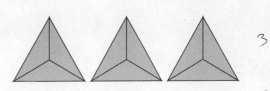 3

6. 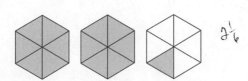 $2\frac{1}{6}$

7. 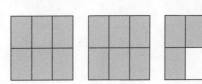 $2\frac{3}{4}$

Shade each figure to show the fraction.

8.
$\frac{5}{8}$

9.
$\frac{1}{12}$

10.
two thirds

Reduce each fraction to lowest terms. If it is in lowest terms, write LT.

11. $\frac{21}{28} =$

12. $\frac{19}{30} =$

13. $\frac{12}{28} =$

14. $\frac{34}{51} =$

15. $\frac{36}{63} =$

16. $\frac{6}{8} =$

17. $\frac{2}{5} =$

18. $\frac{14}{21} =$

19. $\frac{20}{30} =$

20. $\frac{150}{300} =$

Raise each fraction to higher terms using the given denominator.

21. $\dfrac{2}{3} = \dfrac{}{6}$

22. $\dfrac{1}{8} = \dfrac{}{24}$

23. $\dfrac{2}{9} = \dfrac{}{18}$

24. $\dfrac{5}{7} = \dfrac{}{28}$

25. $\dfrac{1}{2} = \dfrac{}{10}$

26. $\dfrac{3}{5} = \dfrac{}{15}$

27. $\dfrac{3}{4} = \dfrac{}{20}$

28. $\dfrac{9}{10} = \dfrac{}{30}$

Compare. Write >, <, or =.

29. $\dfrac{1}{2} \square \dfrac{2}{5}$

30. $\dfrac{2}{3} \square \dfrac{15}{20}$

31. $\dfrac{4}{7} \square \dfrac{7}{10}$

32. $4\dfrac{1}{9} \square 4\dfrac{1}{3}$

33. $\dfrac{5}{8} \square \dfrac{10}{16}$

34. $7\dfrac{3}{9} \square 7\dfrac{1}{4}$

35. $\dfrac{4}{10} \square \dfrac{2}{4}$

36. $\dfrac{6}{8} \square \dfrac{3}{4}$

Change each fraction to a whole or mixed number.

37. $\dfrac{3}{2} =$

38. $\dfrac{15}{4} =$

39. $\dfrac{24}{6} =$

40. $\dfrac{9}{6} =$

41. $\dfrac{45}{7} =$

42. $\dfrac{71}{8} =$

43. $\dfrac{29}{9} =$

44. $\dfrac{12}{3} =$

Change each mixed number to an improper fraction.

45. $4\dfrac{1}{6} =$

46. $5\dfrac{2}{5} =$

47. $13\dfrac{1}{3} =$

48. $1\dfrac{5}{8} =$

Below is a list of the problems in this review and the pages on which the skills are taught. If you missed any problems, turn to the pages listed and practice the skills. Then correct the problems you missed in the Unit Review.

Problems	Pages	Problems	Pages
1-3	11-12	21-28	19-20
4-7	31-33	29-36	23-25, 28
8-10	11	37-44	27, 34
11-20	15-16	45-48	35-36

Unit 2 Adding Fractions

Addition problems with fractions usually ask you to figure out a total amount or a sum. You may have needed to add fractions when using a ruler, doing home repair projects, or figuring out the number of hours that you worked in a week.

In this unit you will apply your understanding of fractions. You will learn how to add fractions, how to solve word problems using fractions, and how to reduce to lowest terms.

Getting Ready

You should be familiar with the skills on this page and the next before you begin this unit. To check your answers, turn to page 170.

 To compare fractions, make sure that the denominators are the same. Then compare the numerators.

Compare the fractions. Write <, >, or =.

1. $\dfrac{1}{3}$ < $\dfrac{2}{3}$
2. $\dfrac{7}{10}$ ☐ $\dfrac{3}{5}$
3. $\dfrac{9}{12}$ ☐ $\dfrac{5}{6}$
4. $\dfrac{2}{3}$ ☐ $\dfrac{3}{4}$

5. $\dfrac{2}{9}$ ☐ $\dfrac{1}{3}$
6. $\dfrac{3}{8}$ ☐ $\dfrac{6}{16}$
7. $\dfrac{8}{11}$ ☐ $\dfrac{5}{11}$
8. $\dfrac{1}{2}$ ☐ $\dfrac{2}{3}$

For review, see Unit 1, pages 23–25.

Getting Ready

 To reduce a fraction to lowest terms, divide the numerator and the denominator by the same number.

Write each fraction in lowest terms.

9. $\dfrac{8}{12} = \dfrac{2}{3}$
10. $\dfrac{15}{20} =$
11. $\dfrac{18}{36} =$
12. $3\dfrac{21}{28} =$
13. $1\dfrac{18}{27} =$

14. $\dfrac{4}{6} =$
15. $\dfrac{8}{10} =$
16. $4\dfrac{14}{28} =$
17. $5\dfrac{2}{4} =$
18. $10\dfrac{9}{12} =$

For review, see Unit 1, pages 15–16.

 To raise a fraction to higher terms, multiply both the numerator and the denominator by the same number.

Write each fraction in higher terms with a denominator of 12.

19. $\dfrac{2}{3} = \dfrac{8}{12}$
20. $\dfrac{3}{4} = \dfrac{}{12}$
21. $\dfrac{1}{2} = \dfrac{}{12}$
22. $2\dfrac{5}{6} = 2\dfrac{}{12}$
23. $5\dfrac{1}{4} = 5\dfrac{}{12}$

24. $\dfrac{5}{6} = \dfrac{}{12}$
25. $\dfrac{1}{3} = \dfrac{}{12}$
26. $9\dfrac{1}{2} = 9\dfrac{}{12}$
27. $3\dfrac{4}{6} = 3\dfrac{}{12}$
28. $1\dfrac{2}{3} = 1\dfrac{}{12}$

For review, see Unit 1, pages 19–20.

 To change an improper fraction to a whole number or a mixed number, divide the numerator by the denominator. Write any remainder as a fraction.

Change each improper fraction to a whole number or a mixed number. Reduce if possible.

29. $\dfrac{11}{5} =$

$$\begin{array}{r} 2\frac{1}{5} \\ 5\overline{)11} \\ -10 \\ \hline 1 \end{array}$$

30. $\dfrac{20}{3} =$
31. $\dfrac{49}{9} =$
32. $\dfrac{30}{10} =$
33. $\dfrac{25}{5} =$

34. $\dfrac{7}{2} =$
35. $\dfrac{8}{5} =$
36. $\dfrac{12}{6} =$
37. $\dfrac{20}{8} =$
38. $\dfrac{15}{12} =$

For review, see Unit 1, pages 33–34.

Adding Fractions with the Same Denominator

To add fractions with the same denominator, add only the numerators. Keep the same denominator.

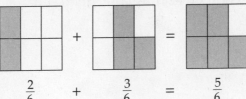

$$\frac{2}{6} + \frac{3}{6} = \frac{5}{6}$$

Use These Steps

Add $\frac{2}{8} + \frac{1}{8}$

1. The denominators are the same. Write the denominator under the fraction bar.

   $\frac{2}{8} + \frac{1}{8} = \frac{}{8}$

2. Add only the numerators. 2 + 1 = 3. Write the sum over the denominator.

   $\frac{2}{8} + \frac{1}{8} = \frac{3}{8}$

Add the fractions and shade in the figures to show the sum.

1.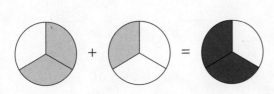

   $\frac{1}{3} + \frac{1}{3} = \frac{2}{3}$

2.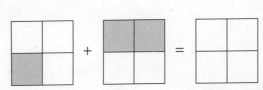

   $\frac{1}{4} + \frac{2}{4} =$

3.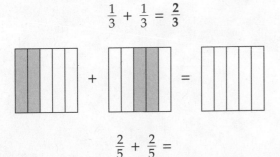

   $\frac{2}{5} + \frac{2}{5} =$

4.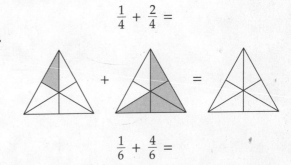

   $\frac{1}{6} + \frac{4}{6} =$

Shade in the figures to show the fractions. Add the fractions. Then shade in the final figure to show the sum.

5.

   $\frac{1}{8} + \frac{2}{8} = \frac{3}{8}$

6.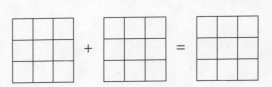

   $\frac{2}{9} + \frac{5}{9} =$

41

Adding Fractions with the Same Denominator

You can add fractions that are lined up vertically as well as horizontally. Remember, when adding fractions with the same denominator, add only the numerators.

Use These Steps

Add $\quad \dfrac{1}{5}$
$\quad\quad +\dfrac{3}{5}$

1. The denominators are the same. Write the denominator under the fraction bar.

$$\dfrac{1}{5} \\ +\dfrac{3}{5} \\ \overline{5}$$

2. Add only the numerators. $1 + 3 = 4$. Write the sum over the denominator.

$$\dfrac{1}{5} \\ +\dfrac{3}{5} \\ \dfrac{4}{5}$$

Add.

1. $\dfrac{3}{6} + \dfrac{2}{6} = \dfrac{5}{6}$

2. $\dfrac{6}{12} + \dfrac{1}{12} =$

3. $\dfrac{3}{15} + \dfrac{10}{15} =$

4. $\dfrac{7}{20} + \dfrac{4}{20} =$

5. $\dfrac{5}{10} + \dfrac{4}{10} =$

6. $\dfrac{6}{14} + \dfrac{3}{14} + \dfrac{4}{14} = \dfrac{13}{14}$

7. $\dfrac{10}{35} + \dfrac{15}{35} + \dfrac{6}{35} =$

8. $\dfrac{7}{40} + \dfrac{3}{40} + \dfrac{11}{40} =$

9. $\dfrac{18}{50} + \dfrac{8}{50} + \dfrac{5}{50} =$

10. $\dfrac{3}{100} + \dfrac{10}{100} + \dfrac{20}{100} =$

11. $\dfrac{6}{25} + \dfrac{7}{25} + \dfrac{1}{25} =$

12. $\dfrac{1}{9} + \dfrac{4}{9} + \dfrac{2}{9} =$

13. $\dfrac{2}{7} + \dfrac{1}{7} + \dfrac{2}{7} =$

14. Before lunch Jenny finished $\dfrac{2}{5}$ of her work. After lunch she finished another $\dfrac{2}{5}$. What part of her work did Jenny finish all together?

    Answer _____

15. Roberto mixed $\dfrac{2}{8}$ gallon of blue paint and $\dfrac{3}{8}$ gallon of white paint. How much paint did he use in all?

    Answer _____

Real-Life Application

On the Job

Felix works at a health club as a trainer. He keeps track of the time each member of the club works on each piece of exercise equipment. On Saturday he worked with Ann, Petra, and Otis. The chart shows how much time each person spent on each different piece of equipment. The time is kept in $\frac{1}{4}$ hours.

Member	Exercise Bike	Treadmill	Rowing Machine
Ann	$\frac{2}{4}$	$\frac{1}{4}$	$\frac{1}{4}$
Petra	$\frac{1}{4}$	$\frac{1}{4}$	0
Otis	$\frac{3}{4}$	$\frac{2}{4}$	$\frac{1}{4}$

Example How much time did Petra spend on the exercise bike and the treadmill on Saturday?

Petra spent $\frac{1}{4}$ hour on the exercise bike and $\frac{1}{4}$ hour on the treadmill.

$$\frac{1}{4} + \frac{1}{4} = \frac{2}{4} = \frac{1}{2}$$

Petra spent $\frac{1}{2}$ hour on the two pieces of equipment.

Use the chart above to answer the following questions. Change improper fractions to whole or mixed numbers. Reduce if possible.

1. How much time all together did Ann, Petra, and Otis spend on the rowing machine on Saturday?

   Answer_____

2. How much time did Ann spend on all three pieces of equipment?

   Answer_____

3. How much total time did Otis spend on the three pieces of equipment?

   Answer_____

4. How much time in all did Otis and Ann spend on the treadmill?

   Answer_____

Adding and Reducing Fractions

When adding fractions, you may get an answer that can be reduced. Always reduce answers to lowest terms.

Use These Steps

Add $\frac{1}{8} + \frac{3}{8}$

1. The denominators are the same. Write the denominator under the fraction bar.

    $\frac{1}{8} + \frac{3}{8} = \frac{}{8}$

2. Add the numerators.

    $\frac{1}{8} + \frac{3}{8} = \frac{4}{8}$

3. Reduce to lowest terms.

    $\frac{4}{8} = \frac{4 \div 4}{8 \div 4} = \frac{1}{2}$

Add. Reduce to lowest terms.

1. $\frac{1}{4} + \frac{1}{4} = \frac{2}{4} = \frac{2 \div 2}{4 \div 2} = \frac{1}{2}$

2. $\frac{3}{6} + \frac{1}{6} =$

3. $\frac{6}{10} + \frac{2}{10} =$

4. $\frac{7}{12} + \frac{1}{12} =$

5. $\frac{4}{15} + \frac{6}{15} =$

6. $\frac{9}{45} + \frac{18}{45} =$

7. $\frac{1}{9} + \frac{1}{9} + \frac{4}{9} =$

8. $\frac{1}{16} + \frac{2}{16} + \frac{1}{16} =$

9. $\frac{22}{44} + \frac{11}{44} + \frac{1}{44} =$

10. $\frac{9}{16} + \frac{5}{16}$

11. $\frac{8}{20} + \frac{10}{20}$

12. $\frac{7}{18} + \frac{7}{18}$

13. $\frac{8}{24} + \frac{12}{24}$

14. $\frac{3}{21} + \frac{4}{21}$

15. $\frac{12}{30} + \frac{5}{30} + \frac{3}{30}$

16. $\frac{15}{36} + \frac{11}{36} + \frac{4}{36}$

17. $\frac{12}{25} + \frac{3}{25} + \frac{5}{25}$

18. $\frac{16}{64} + \frac{20}{64} + \frac{12}{64}$

19. $\frac{20}{50} + \frac{10}{50} + \frac{5}{50}$

20. It is $\frac{2}{10}$ mile from the stop sign to the gas station. It is $\frac{6}{10}$ mile from the gas station to the traffic signal. How far is it from the stop sign to the traffic signal?

    Answer _____

21. Julia is sewing two ribbons side by side. One is $\frac{2}{16}$ inch wide. The other is $\frac{6}{16}$ inch wide. How wide are they all together?

    Answer _____

44

Adding Fractions and Changing to Whole Numbers

Sometimes the answer to an addition problem is an improper fraction that can be changed to a whole number.

Use These Steps

Add $\frac{4}{9} + \frac{5}{9}$

1. The denominators are the same. Write the denominator under the fraction bar.

    $\frac{4}{9} + \frac{5}{9} = \frac{}{9}$

2. Add the numerators.

    $\frac{4}{9} + \frac{5}{9} = \frac{9}{9}$

3. The answer is an improper fraction. Divide the numerator by the denominator to get a whole number.

    $\frac{9}{9} = 9\overline{)9}^{\,1} = 1$

Add. Change each sum to a whole number.

1. $\frac{2}{3} + \frac{1}{3} = \frac{3}{3}$

    $\frac{3}{3} = 3\overline{)3}^{\,1} = 1$

2. $\frac{5}{6} + \frac{1}{6} =$

3. $\frac{5}{8} + \frac{3}{8} =$

4. $\frac{1}{4} + \frac{3}{4} =$

5. $\frac{12}{5} + \frac{3}{5} = \frac{15}{5}$

    $\frac{15}{5} = 5\overline{)15}^{\,3} = 3$

6. $\frac{6}{4} + \frac{2}{4} =$

7. $\frac{4}{3} + \frac{5}{3} =$

8. $\frac{7}{10} + \frac{13}{10} =$

9. $\frac{4}{3} + \frac{1}{3} + \frac{1}{3} =$

10. $\frac{7}{4} + \frac{2}{4} + \frac{3}{4} =$

11. $\frac{25}{15} + \frac{15}{15} + \frac{5}{15} =$

12. $\frac{3}{8} + \frac{5}{8}$

13. $\frac{7}{9} + \frac{20}{9}$

14. $\frac{12}{8} + \frac{20}{8}$

15. $\frac{2}{15} + \frac{13}{15}$

16. $\frac{5}{2} + \frac{5}{2}$

17. Casey welded three lengths of pipe together: two that were $\frac{9}{12}$ foot long and one that was $\frac{6}{12}$ foot long. How long is the new pipe?

    Answer _____

18. It rained three days last week in Centerville. On Monday it rained $\frac{2}{10}$ inch. On Wednesday $\frac{1}{10}$ inch rain fell. It rained $\frac{7}{10}$ inch on Friday. What was the total rainfall last week?

    Answer _____

Adding Fractions and Changing to Mixed Numbers

Sometimes the answer to an addition problem is an improper fraction that can be changed to a mixed number.

Use These Steps

Add $\frac{3}{4} + \frac{3}{4}$

1. The denominators are the same. Add the numerators.

   $\frac{3}{4} + \frac{3}{4} = \frac{6}{4}$

2. The answer is an improper fraction. Divide the numerator by the denominator to get a mixed number. Write the remainder as a fraction.

   $4\overline{)6}1\frac{2}{4}$
   $\underline{-4}$
   2

3. Reduce the fraction to lowest terms.

   $1\frac{2}{4} = 1\frac{2 \div 2}{4 \div 2} = 1\frac{1}{2}$

Add. Change each sum to a mixed number. Reduce if possible.

1. $\frac{4}{6} + \frac{4}{6} = \frac{8}{6}$
   $\frac{8}{6} = 1\frac{2}{6} = 1\frac{1}{3}$

2. $\frac{6}{4} + \frac{8}{4} =$

3. $\frac{13}{10} + \frac{1}{10} =$

4. $\frac{8}{9} + \frac{7}{9} =$

5. $\frac{4}{5} + \frac{3}{5} + \frac{4}{5} = \frac{11}{5}$
   $\frac{11}{5} = 2\frac{1}{5}$

6. $\frac{11}{3} + \frac{1}{3} + \frac{1}{3} =$

7. $\frac{4}{9} + \frac{5}{9} + \frac{8}{9} =$

8. $\frac{3}{4}$
   $+\frac{7}{4}$

9. $\frac{9}{15}$
   $\frac{11}{15}$
   $+\frac{3}{15}$

10. $\frac{16}{20}$
    $+\frac{7}{20}$

11. $\frac{15}{35}$
    $\frac{25}{35}$
    $+\frac{10}{35}$

12. $\frac{33}{50}$
    $+\frac{27}{50}$

13. Anita rode her bike in the park for $\frac{3}{4}$ hour. Then she ate lunch for $\frac{3}{4}$ hour. How much time did she spend in the park?

    Answer _____

14. On Fortune Highway, it is $\frac{8}{10}$ mile from Adams Street to Bailey Street. The distance from Bailey Street to Carter Street is $\frac{9}{10}$ mile. How far is it from Adams Street to Carter Street?

    Answer _____

Mixed Review

Reduce each fraction to lowest terms. If it is in lowest terms, write LT.

1. $\frac{7}{14} =$
2. $\frac{9}{3} =$
3. $4\frac{4}{12} =$
4. $\frac{5}{20} =$
5. $\frac{15}{4} =$

6. $\frac{25}{2} =$
7. $3\frac{9}{36} =$
8. $\frac{6}{6} =$
9. $\frac{3}{21} =$
10. $15\frac{2}{7} =$

Add. Change any improper fractions to whole or mixed numbers. Reduce fractions to lowest terms.

11. $\frac{5}{7} + \frac{6}{7} =$
12. $\frac{4}{11} + \frac{15}{11} =$
13. $\frac{13}{15} + \frac{5}{15} =$
14. $\frac{5}{6} + \frac{2}{6} =$

15. $\frac{1}{5} + \frac{1}{5} =$
16. $\frac{7}{10} + \frac{2}{10} =$
17. $\frac{14}{25} + \frac{8}{25} =$
18. $\frac{55}{100} + \frac{12}{100} + \frac{4}{100} =$

19. $\frac{2}{10} + \frac{6}{10} =$
20. $\frac{1}{9} + \frac{5}{9} =$
21. $\frac{12}{30} + \frac{13}{30} =$
22. $\frac{3}{15} + \frac{7}{15} + \frac{2}{15} =$

23. $\frac{25}{100} + \frac{75}{100} =$
24. $\frac{5}{15} + \frac{25}{15} =$
25. $\frac{50}{50} + \frac{100}{50} =$
26. $\frac{2}{6} + \frac{2}{6} + \frac{2}{6} =$

27. $\frac{8}{9} + \frac{8}{9} =$
28. $\frac{35}{60} + \frac{45}{60} =$
29. $\frac{21}{25} + \frac{40}{25} =$
30. $\frac{16}{24} + \frac{10}{24} + \frac{2}{24} =$

31. $\frac{3}{8}$
    $+\frac{4}{8}$

32. $\frac{5}{7}$
    $\frac{3}{7}$
    $+\frac{4}{7}$

33. $\frac{5}{9}$
    $+\frac{1}{9}$

34. $\frac{7}{10}$
    $+\frac{7}{10}$

35. $\frac{8}{15}$
    $\frac{10}{15}$
    $+\frac{2}{15}$

Real-Life Application — Time Off

For a report she was writing, Wanling asked 12 people what their favorite sport was. She made a circle graph. Since she asked 12 people, she divided the circle graph into 12 equal parts.

Example Four people said that baseball is their favorite sport. How many sections did Wanling use to show what fraction of the people liked baseball? Write the fraction. Reduce if possible.

Since 4 of the 12 people liked baseball, she used 4 sections.

$$\frac{4}{12} = \frac{1}{3}$$

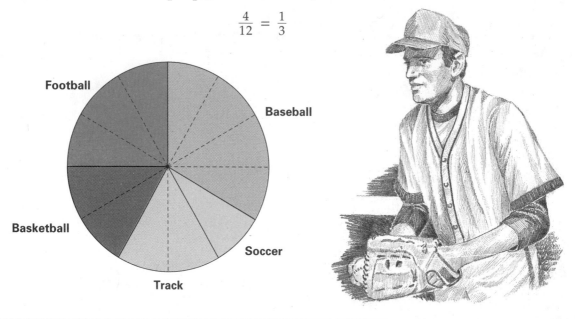

Solve. Reduce if possible.

1. How many sections did she use for football? Write the fraction.

   Answer_____

2. How many sections did she use for baseball and football together? Write the fraction.

   Answer_____

3. How many sections did she use for soccer? Write the fraction.

   Answer_____

4. What fraction shows how many people chose track?

   Answer_____

5. What fraction of the people Wanling spoke to chose football, baseball, and basketball?

   Answer_____

6. If you add the fractions for all of the sports, what is the total?

   Answer_____

Adding Mixed Numbers

You know that a mixed number has a whole number and a fraction. When adding mixed numbers, you need to add the whole numbers together and the fractions together.

Use These Steps

Add $3\frac{2}{9} + 2\frac{1}{9}$

1. Line up the fractions and the whole numbers in columns.

   $3\frac{2}{9}$
   $+2\frac{1}{9}$

2. Add the fractions.

   $3\frac{2}{9}$
   $+2\frac{1}{9}$
   $\frac{3}{9}$

3. Add the whole numbers. Reduce the fraction to lowest terms.

   $3\frac{2}{9}$
   $+2\frac{1}{9}$
   $5\frac{3}{9} = 5\frac{1}{3}$

Add. Reduce if possible.

1. $4\frac{1}{6} + 5\frac{2}{6} =$

   $4\frac{1}{6}$
   $+5\frac{2}{6}$
   $9\frac{3}{6} = 9\frac{1}{2}$

2. $1\frac{1}{4} + 7\frac{1}{4} =$

3. $2\frac{3}{8} + 5\frac{4}{8} =$

4. $6\frac{3}{10} + 3\frac{5}{10} =$

5. $16\frac{15}{20}$
   $+5\frac{4}{20}$

6. $23\frac{13}{25}$
   $+18\frac{7}{25}$

7. $30\frac{2}{10}$
   $+33\frac{5}{10}$

8. $42\frac{20}{50}$
   $+37\frac{10}{50}$

9. $6\frac{3}{12} + 4\frac{1}{12} + 7\frac{2}{12} =$

   $6\frac{3}{12}$
   $4\frac{1}{12}$
   $+7\frac{2}{12}$
   $17\frac{6}{12} = 17\frac{1}{2}$

10. $10\frac{2}{7} + 13\frac{1}{7} + 22\frac{2}{7} =$

11. $3\frac{1}{10} + 5\frac{3}{10} + 8\frac{1}{10} =$

Adding Mixed Numbers

Sometimes when you add mixed numbers, the fraction part of the answer is an improper fraction. When this happens, change the improper fraction to a whole or mixed number. Then add the result to the whole number of the answer.

Use These Steps

Add $3\frac{3}{5} + 2\frac{4}{5}$

1. Line up the mixed numbers. Add the fractions. Add the whole numbers.

    $3\frac{3}{5}$
    $+2\frac{4}{5}$
    $\overline{5\frac{7}{5}}$

2. The fraction part of the answer is an improper fraction. Change it to a mixed number.

    $\frac{7}{5} = 1\frac{2}{5}$

3. Add the result to the whole number part of the answer.

    $5 + 1\frac{2}{5} = 6\frac{2}{5}$

Add. Change any improper fraction to a mixed number. Reduce if possible.

1. $2\frac{1}{6} + 4\frac{5}{6} =$

    $2\frac{1}{6}$
    $+4\frac{5}{6}$
    $\overline{6\frac{6}{6}} = 6 + 1 = 7$

2. $5\frac{3}{4} + 5\frac{3}{4} =$

3. $9\frac{9}{10} + 15\frac{5}{10} =$

4. $7\frac{2}{9} + 19\frac{8}{9} =$

5. $16\frac{3}{5}$
   $14\frac{4}{5}$
   $+9\frac{3}{5}$
   $\overline{}$

6. $29\frac{9}{10}$
   $42\frac{6}{10}$
   $+37\frac{8}{10}$
   $\overline{}$

7. $56\frac{8}{25}$
   $32\frac{19}{25}$
   $+69\frac{23}{25}$
   $\overline{}$

8. $25\frac{50}{100}$
   $13\frac{25}{100}$
   $+7\frac{25}{100}$
   $\overline{}$

9. Phil watched TV for $1\frac{1}{2}$ hours on Monday, $2\frac{1}{2}$ hours on Tuesday, and $1\frac{1}{2}$ hours on Wednesday. How many hours did he watch TV all together?

10. Jana drove $3\frac{7}{10}$ miles from work to the supermarket, $1\frac{3}{10}$ miles to the hardware store, and $6\frac{1}{10}$ miles back home. How far did she drive in all?

Answer _____

Answer _____

Adding Mixed Numbers, Whole Numbers, and Fractions

When adding mixed numbers, whole numbers, and fractions, be sure to add all of the whole numbers together and all of the fractions together. Reduce answers if possible.

Use These Steps

Add $3\frac{3}{4} + \frac{3}{4} + 6$

1. Add the fractions. Add the whole numbers.

$$\begin{array}{r} 3\frac{3}{4} \\ \frac{3}{4} \\ +6 \\ \hline 9\frac{6}{4} \end{array}$$

2. Change the improper fraction to a mixed number.

$$\frac{6}{4} = 1\frac{2}{4}$$

3. Add the result to the whole number part of the sum. Reduce.

$$9 + 1\frac{2}{4} = 10\frac{2}{4} = 10\frac{1}{2}$$

Add. Reduce if possible.

1.
$$\begin{array}{r} 2\frac{3}{8} \\ \frac{7}{8} \\ +4 \\ \hline 6\frac{10}{8} \end{array} = 6 + 1\frac{2}{8} = 7\frac{2}{8} = 7\frac{1}{4}$$

2.
$$\begin{array}{r} 3\frac{2}{5} \\ 7 \\ +\frac{4}{5} \\ \hline \end{array}$$

3.
$$\begin{array}{r} \frac{6}{10} \\ 9\frac{7}{10} \\ +6 \\ \hline \end{array}$$

4. $\frac{15}{16} + 14\frac{13}{16} + \frac{4}{16} =$

5. $17\frac{8}{9} + \frac{4}{9} + 16\frac{2}{9} =$

6. $2\frac{7}{12} + 8\frac{8}{12} + 7 =$

7. Lenore needs $3\frac{1}{8}$ yards of material for a jacket, 4 yards for a dress, and $\frac{5}{8}$ yard for a scarf. How many yards of material does she need in all?

8. Lenore needs $2\frac{3}{8}$ yards, 2 yards, and $\frac{3}{8}$ yard of material to make a matching outfit for her daughter. How many yards in all will she need for her daughter's outfit?

Answer _____   Answer _____

Mixed Review

Reduce each fraction to lowest terms. Write LT if the fraction is in lowest terms.

1. $\dfrac{7}{8} =$
2. $\dfrac{9}{2} =$
3. $2\dfrac{12}{4} =$
4. $6\dfrac{3}{5} =$
5. $1\dfrac{1}{3} =$

6. $\dfrac{15}{20} =$
7. $7\dfrac{5}{4} =$
8. $3\dfrac{2}{3} =$
9. $\dfrac{21}{7} =$
10. $\dfrac{2}{2} =$

Add. Reduce if possible.

11. $\dfrac{2}{9} + \dfrac{3}{9} =$
12. $\dfrac{5}{8} + \dfrac{1}{8} =$
13. $\dfrac{4}{5} + \dfrac{3}{5} =$
14. $\dfrac{9}{14} + \dfrac{7}{14} + \dfrac{3}{14} =$

15. $8\dfrac{3}{8} + 3\dfrac{7}{8} =$
16. $5\dfrac{7}{10} + 4\dfrac{3}{10} =$
17. $2\dfrac{7}{12} + 8\dfrac{8}{12} =$
18. $19\dfrac{5}{6} + 12 + 1\dfrac{4}{6} =$

19. $10\dfrac{5}{16}$
    $3\dfrac{2}{16}$
    $+\ \dfrac{7}{16}$

20. $4\dfrac{1}{2}$
    $3\dfrac{1}{2}$
    $+9\dfrac{1}{2}$

21. $\dfrac{3}{4}$
    $\dfrac{3}{4}$
    $+\dfrac{1}{4}$

22. 5
    $7\dfrac{7}{8}$
    $+12\dfrac{6}{8}$

23. $23\dfrac{10}{25}$
    $15\dfrac{24}{25}$
    $+32\dfrac{20}{25}$

24. Alma is $62\dfrac{3}{4}$ inches tall. Her sister is $\dfrac{3}{4}$ inch taller. How tall is her sister?

25. Alma's mother is $1\dfrac{1}{4}$ inches taller than Alma. How tall is her mother?

Answer _____

Answer _____

Problem Solving: Using a Ruler

The ruler below is divided into inches and fractions of an inch: halves, fourths, eighths, and sixteenths.

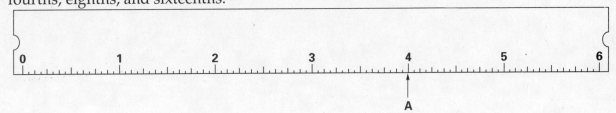

To find the sum of two measures on the ruler, first add the measures. Then find the answer on the ruler.

Example Find the sum of $1\frac{1}{2}$ inches and $2\frac{1}{2}$ inches on the ruler.

▶ **Step 1.** Add the measures.

$$\begin{array}{r} 1\frac{1}{2} \\ +2\frac{1}{2} \\ \hline 3\frac{2}{2} \end{array} = 3 + 1 = 4 \text{ inches}$$

▶ **Step 2.** Find 4 inches on the ruler. Draw an arrow pointing to the answer and label it A.

Find each sum on the ruler above.

1. Find the sum of $\frac{1}{4}$ inch and $1\frac{1}{2}$ inches. Draw an arrow pointing to the answer and label it B.

2. Find the sum of $1\frac{5}{8}$ inches and 2 inches. Draw an arrow pointing to the answer and label it C.

Answer_____   Answer_____

3. Find 3 inches on the ruler. Add $2\frac{3}{4}$ inches more. Draw an arrow pointing to the answer and label it D.

4. Find $4\frac{1}{2}$ inches on the ruler. Add $1\frac{1}{16}$ inches more. Draw an arrow pointing to the answer and label it E.

Answer_____   Answer_____

Find each sum on the ruler below.

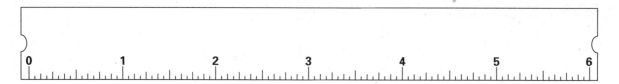

5. Find the sum of $\frac{3}{8}$ inch and $3\frac{3}{8}$ inches on the ruler. Draw an arrow and label it A.

   Answer _____

6. Find the sum of 4 inches and $\frac{1}{8}$ inch on the ruler. Draw an arrow and label it B.

   Answer _____

7. Find the sum of $1\frac{5}{16}$ inches and $2\frac{3}{16}$ inches on the ruler. Draw an arrow and label it C.

   Answer _____

8. Find the sum of $3\frac{1}{8}$ inches and $1\frac{7}{16}$ inches on the ruler. Draw an arrow and label it D.

   Answer _____

9. Find the sum of $\frac{7}{8}$ inch and $1\frac{7}{8}$ inches on the ruler. Draw an arrow and label it E.

   Answer _____

10. Find the sum of $2\frac{1}{2}$ inches and $2\frac{3}{4}$ inches on the ruler. Draw an arrow and label it F.

    Answer _____

11. Find $1\frac{5}{8}$ inches on the ruler. Add $2\frac{3}{4}$ inches more. Draw an arrow and label it G.

    Answer _____

12. Find $4\frac{9}{16}$ inches on the ruler. Add $1\frac{7}{16}$ inches more. Draw an arrow and label it H.

    Answer _____

Finding Common Denominators

When you have two fractions with different denominators, you may need to change the fractions to equal fractions in higher terms with a common denominator.

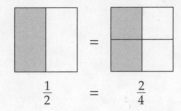

$$\frac{1}{2} = \frac{2}{4}$$

Use These Steps

Write $\frac{1}{3}$ and $\frac{5}{6}$ with common denominators.

1. Compare the denominators. 3 divides evenly into 6. The larger number, 6, works as the common denominator.

    $\frac{1}{3} = \frac{}{6}$
    $\frac{5}{6} = \frac{}{6}$

2. Change $\frac{1}{3}$ to a fraction in higher terms with 6 as the denominator.

    $\frac{1}{3} = \frac{2}{6}$
    $\frac{5}{6} = \frac{5}{6}$

Write each set of fractions with common denominators.

1. $\frac{1}{2}$ and $\frac{3}{10}$

    $\frac{1}{2} = \frac{5}{10}$
    $\frac{3}{10} = \frac{3}{10}$

2. $\frac{4}{5}$ and $\frac{7}{15}$

3. $\frac{2}{3}$ and $\frac{4}{9}$

4. $\frac{7}{12}$ and $\frac{5}{6}$

5. $\frac{3}{4}$ and $\frac{5}{8}$

6. $\frac{1}{2}$ and $\frac{5}{6}$

7. $\frac{7}{21}$ and $\frac{3}{7}$

8. $\frac{1}{5}$ and $\frac{3}{25}$

9. $\frac{3}{14}$ and $\frac{2}{7}$

10. $\frac{5}{8}$ and $\frac{7}{24}$

11. $\frac{9}{20}$ and $\frac{3}{10}$

12. $\frac{2}{9}$ and $\frac{5}{27}$

13. $\frac{1}{2}$, $\frac{3}{4}$, and $\frac{7}{8}$

    $\frac{1}{2} = \frac{4}{8}$
    $\frac{3}{4} = \frac{6}{8}$
    $\frac{7}{8} = \frac{7}{8}$

14. $\frac{2}{3}$, $\frac{5}{6}$, and $\frac{1}{18}$

15. $\frac{1}{3}$, $\frac{3}{5}$, and $\frac{7}{15}$

Adding Fractions with Different Denominators

When fractions in an addition problem have different denominators, you must first change the fractions to equal fractions with common denominators. Then add the numerators.

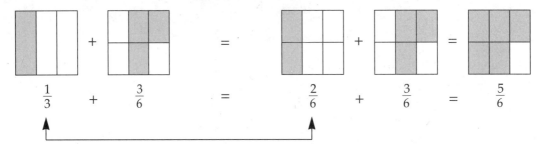

$$\frac{1}{3} + \frac{3}{6} = \frac{2}{6} + \frac{3}{6} = \frac{5}{6}$$

Use These Steps

Add $\frac{1}{4} + \frac{3}{8}$

1. Compare the denominators. 4 will divide evenly into 8. 8 is the common denominator. Change 1/4 to higher terms with 8 as the common denominator.

$$\frac{1}{4} = \frac{1 \times 2}{4 \times 2} = \frac{2}{8}$$

2. The denominators are the same. Add the numerators.

$$\frac{1}{4} = \frac{2}{8}$$
$$+\frac{3}{8} = \frac{3}{8}$$
$$\overline{\frac{5}{8}}$$

Find common denominators. Change the fractions to higher terms. Then add.

1. $\frac{1}{5} + \frac{5}{10} =$

   $\frac{1}{5} = \frac{2}{10}$
   $+\frac{5}{10} = \frac{5}{10}$
   $\overline{\frac{7}{10}}$

2. $\frac{2}{6} + \frac{1}{2} =$

3. $\frac{3}{14} + \frac{1}{7} =$

4. $\frac{1}{4} + \frac{1}{2} =$

5. $\frac{5}{8}$
   $+\frac{1}{4}$

6. $\frac{1}{3}$
   $+\frac{5}{9}$

7. $\frac{3}{10}$
   $+\frac{2}{5}$

8. $\frac{1}{15}$
   $+\frac{1}{5}$

9. $\frac{1}{10}$
   $+\frac{5}{20}$

Adding Fractions with Different Denominators

When you add fractions that have different denominators, you will need to find a common denominator. This means raising one or more of the fractions to higher terms.

Use These Steps

Add $\frac{3}{4} + \frac{1}{8} + \frac{1}{2}$.

1. Change $\frac{3}{4}$ to higher terms with 8 as the denominator.

    $\frac{3}{4} = \frac{3 \times 2}{4 \times 2} = \frac{6}{8}$

2. Change $\frac{1}{2}$ to higher terms with 8 as the denominator.

    $\frac{1}{2} = \frac{1 \times 4}{2 \times 4} = \frac{4}{8}$

3. Add the fractions. Change the answer to a mixed number.

    $\frac{3}{4} = \frac{6}{8}$
    $\frac{1}{8} = \frac{1}{8}$
    $+\frac{1}{2} = \frac{4}{8}$
    $\overline{\frac{11}{8} = 1\frac{3}{8}}$

Find common denominators. Then add. Reduce if possible.

1. $\frac{3}{10} + \frac{2}{5} + \frac{1}{2} =$

    $\frac{3}{10} = \frac{3}{10}$
    $\frac{2}{5} = \frac{4}{10}$
    $+\frac{1}{2} = \frac{5}{10}$
    $\overline{\frac{12}{10} = 1\frac{2}{10} = 1\frac{1}{5}}$

2. $\frac{1}{4} + \frac{1}{8} + \frac{3}{16} =$

3. $\frac{5}{18} + \frac{3}{9} + \frac{2}{3} =$

4. $\frac{1}{9} + \frac{1}{3} + \frac{5}{18} =$

5. $\frac{1}{4} + \frac{2}{8} + \frac{5}{16} =$

6. $\frac{2}{10} + \frac{1}{5} + \frac{1}{2} =$

7. $\frac{6}{20}$
    $\frac{2}{5}$
    $+\frac{1}{4}$

8. $\frac{5}{45}$
    $\frac{1}{9}$
    $+\frac{1}{5}$

9. $\frac{5}{20}$
    $\frac{4}{15}$
    $+\frac{10}{60}$

10. $\frac{1}{4}$
    $\frac{10}{52}$
    $+\frac{1}{13}$

11. $\frac{9}{25}$
    $\frac{17}{50}$
    $+\frac{29}{100}$

Adding Fractions with Different Denominators

When you add fractions with different denominators, first find a common denominator. Then add. Reduce if possible.

Use These Steps

Add $\frac{1}{7} + \frac{4}{14}$

1. Change $\frac{1}{7}$ to higher terms with 14 as the common denominator.

   $\frac{1}{7} = \frac{1 \times 2}{7 \times 2} = \frac{2}{14}$

2. Add the fractions. Reduce.

   $\frac{1}{7} = \frac{2}{14}$
   $+ \frac{4}{14} = \frac{4}{14}$
   $\frac{6}{14} = \frac{3}{7}$

Add. Reduce if possible.

1. $\frac{3}{14} + \frac{2}{7} =$

   $\frac{3}{14} = \frac{3}{14}$
   $+\frac{2}{7} = \frac{4}{14}$
   $\frac{7}{14} = \frac{1}{2}$

2. $\frac{2}{15} + \frac{1}{5} =$

3. $\frac{1}{4} + \frac{3}{20} =$

4. $\frac{14}{25} + \frac{2}{5} =$

5. $\frac{2}{9}$
   $+\frac{5}{18}$

6. $\frac{1}{3}$
   $+\frac{6}{12}$

7. $\frac{3}{10}$
   $+\frac{1}{2}$

8. $\frac{21}{50}$
   $+\frac{6}{25}$

9. $\frac{1}{10}$
   $+\frac{15}{100}$

10. $\frac{1}{4} + \frac{1}{2} + \frac{1}{8} =$

11. $\frac{2}{10} + \frac{1}{2} + \frac{1}{5} =$

12. $\frac{1}{3} + \frac{2}{6} + \frac{2}{12} =$

13. Joan made cinnamon rolls. She used $\frac{1}{2}$ cup sugar for the rolls and $\frac{1}{4}$ cup sugar for the frosting. How much sugar did she use in all?

14. Chris spent $\frac{1}{6}$ of his allowance on baseball cards and $\frac{1}{3}$ on a magazine. What fraction of his allowance did he spend in all?

Answer _____

Answer _____

Adding Fractions with Different Denominators

When the sum of two or more fractions is an improper fraction, be sure to change the sum to a mixed or whole number.

Use These Steps

Add $\frac{2}{3} + \frac{5}{9}$

1. Change $\frac{2}{3}$ to higher terms with 9 as the common denominator.

$$\frac{2}{3} = \frac{2 \times 3}{3 \times 3} = \frac{6}{9}$$

2. Add the fractions. Change the sum to a mixed number.

$$\frac{2}{3} = \frac{6}{9}$$
$$+\frac{5}{9} = \frac{5}{9}$$
$$\frac{11}{9} = 1 + \frac{2}{9} = 1\frac{2}{9}$$

Add. Change improper fractions to mixed or whole numbers. Reduce if possible.

1. $\frac{5}{6} + \frac{1}{3} =$

$\frac{5}{6} = \frac{5}{6}$
$+\frac{1}{3} = \frac{2}{6}$
$\frac{7}{6} = 1 + \frac{1}{6} = 1\frac{1}{6}$

2. $\frac{7}{8} + \frac{3}{4} =$

3. $\frac{5}{6} + \frac{1}{2} =$

4. $\frac{3}{8} + \frac{15}{16} =$

5. $\frac{1}{4} + \frac{1}{3} + \frac{7}{12} =$

6. $\frac{7}{10} + \frac{1}{2} + \frac{4}{5} =$

7. $\frac{9}{24} + \frac{5}{8} + \frac{2}{3} =$

8. $\frac{5}{8}$
$+\frac{3}{4}$

9. $\frac{5}{7}$
$+\frac{9}{14}$

10. $\frac{11}{12}$
$+\frac{2}{3}$

11. $\frac{4}{5}$
$+\frac{11}{15}$

12. $\frac{8}{10}$
$+\frac{10}{50}$

13. What is the total weight of three boxes of candy if they weigh $\frac{12}{16}$ pound, $\frac{3}{4}$ pound, and $\frac{6}{8}$ pound?

14. Find the combined lengths of three pieces of ribbon that measure $\frac{4}{5}$ yard, $\frac{3}{10}$ yard, and $\frac{1}{2}$ yard.

Answer _____

Answer _____

Mixed Review

Raise each fraction to higher terms.

1. $\frac{1}{4} = \frac{}{8}$
2. $\frac{3}{4} = \frac{}{12}$
3. $\frac{2}{3} = \frac{}{9}$
4. $\frac{5}{6} = \frac{}{18}$
5. $\frac{3}{5} = \frac{}{20}$

Reduce each fraction to lowest terms. Write LT if in lowest terms.

6. $\frac{6}{9} =$
7. $\frac{6}{10} =$
8. $\frac{3}{18} =$
9. $\frac{13}{25} =$
10. $\frac{8}{16} =$

Change each fraction to a whole or mixed number.

11. $\frac{6}{6} =$
12. $\frac{15}{12} =$
13. $\frac{9}{3} =$
14. $\frac{22}{7} =$
15. $\frac{45}{20} =$

Add. Reduce if possible.

16. $\frac{3}{8} + \frac{1}{8} =$
17. $\frac{7}{12} + \frac{5}{12} =$
18. $\frac{1}{9} + \frac{5}{9} + \frac{7}{9} =$

19. $\frac{2}{5} + \frac{1}{10} =$
20. $\frac{3}{16} + \frac{1}{4} =$
21. $\frac{5}{24} + \frac{1}{6} + \frac{1}{4} =$

22. $\frac{5}{7} + \frac{13}{14} =$
23. $\frac{2}{3} + \frac{11}{12} =$
24. $\frac{13}{18} + \frac{1}{2} + \frac{5}{6} =$

25. $\frac{1}{2}$
    $+ \frac{1}{10}$

26. $2\frac{9}{14}$
    $+ 3\frac{3}{14}$

27. $6\frac{11}{12}$
    $+ 7\frac{11}{12}$

28. $\frac{2}{5}$
    $+ \frac{11}{25}$

29. $3\frac{1}{20}$
    $+ 7\frac{9}{20}$

Real-Life Application

Daily Living

Mona, Theresa, Lisa, and Janet joined a diet club. Each week they write down the number of pounds they have lost. The person who loses the most weight each month wins a movie pass.

Downtown Diet Club					
	Week 1	Week 2	Week 3	Week 4	Total
Mona	$\frac{13}{16}$	$\frac{3}{4}$	$\frac{5}{8}$	$\frac{3}{8}$	$2\frac{9}{16}$
Theresa	$\frac{1}{2}$	$\frac{3}{4}$	$\frac{9}{16}$	$\frac{5}{13}$	
Lisa	$\frac{3}{4}$	$\frac{15}{16}$	$\frac{1}{2}$	$\frac{3}{8}$	
Janet	$\frac{3}{8}$	$\frac{1}{4}$	$\frac{7}{8}$	$\frac{3}{16}$	
Jay					

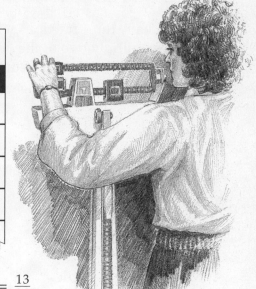

Example How much weight did Mona lose over the four weeks?

$$\frac{13}{16} = \frac{13}{16}$$
$$\frac{3}{4} = \frac{12}{16}$$
$$\frac{5}{8} = \frac{10}{16}$$
$$+\frac{3}{8} = \frac{6}{16}$$
$$\frac{41}{16} = 2\frac{9}{16}$$

Mona lost $2\frac{9}{16}$ pounds.

Solve. Reduce if possible.

1. How much weight did Theresa lose in all?

2. How much weight did Lisa lose in all?

Answer_____ Answer_____

3. How much weight did Janet lose in all?

4. Who won the movie pass?

Answer_____ Answer_____

Comparing Denominators

The common denominator in a fractions problem may not always be the largest denominator. You can also find the common denominator by multiplying the denominators.

When you have three fractions with different denominators, you can sometimes multiply two of the denominators by each other to find the common denominator. Make sure the third denominator will divide evenly into the new denominator.

Use These Steps

Write $\frac{2}{3}$ and $\frac{1}{4}$ with a common denominator.

1. Compare the denominators. Since 3 will not divide evenly into 4, multiply 3 × 4 to get the common denominator.

   $\frac{2}{3} = \frac{}{12}$
   $\frac{1}{4} = \frac{}{12}$

2. Raise the fractions to higher terms with 12 as the denominator.

   $\frac{2}{3} = \frac{8}{12}$
   $\frac{1}{4} = \frac{3}{12}$

Find the common denominator for each set of fractions. Raise each fraction to higher terms.

1. $\frac{1}{5}$ and $\frac{2}{3}$
   $\frac{1}{5} = \frac{3}{15}$
   $\frac{2}{3} = \frac{10}{15}$

2. $\frac{2}{5}$ and $\frac{1}{4}$

3. $\frac{1}{6}$ and $\frac{3}{5}$

4. $\frac{3}{7}$ and $\frac{1}{2}$

5. $\frac{1}{3}$, $\frac{1}{4}$, and $\frac{1}{6}$
   $\frac{1}{3} = \frac{4}{12}$
   $\frac{1}{4} = \frac{3}{12}$
   $\frac{1}{6} = \frac{2}{12}$

6. $\frac{1}{2}$, $\frac{2}{5}$, and $\frac{3}{7}$

7. $\frac{4}{9}$, $\frac{1}{3}$, and $\frac{2}{7}$

8. $\frac{5}{9}$ and $\frac{1}{4}$

9. $\frac{2}{3}$, $\frac{2}{7}$, and $\frac{5}{6}$

10. $\frac{3}{4}$ and $\frac{2}{9}$

11. $\frac{3}{8}$, $\frac{3}{5}$, and $\frac{2}{3}$

Adding Fractions with Different Denominators

When multiplying the denominators to find a common denominator, change all fractions to higher terms. Reduce the answer if possible.

Use These Steps

Add $\frac{1}{6} + \frac{3}{5}$

1. Multiply the denominators to find the common denominator.

$$\frac{1}{6} = \frac{}{30}$$
$$+\frac{3}{5} = \frac{}{30}$$

2. Raise both fractions to higher terms with 30 as the common denominator.

$$\frac{1}{6} = \frac{5}{30}$$
$$+\frac{3}{5} = \frac{18}{30}$$

3. Add.

$$\frac{1}{6} = \frac{5}{30}$$
$$+\frac{3}{5} = \frac{18}{30}$$
$$\frac{23}{30}$$

Add. Reduce if possible.

1. $\frac{1}{2} + \frac{1}{3} =$

$$\frac{1}{2} = \frac{3}{6}$$
$$+\frac{1}{3} = \frac{2}{6}$$
$$\frac{5}{6}$$

2. $\frac{2}{6} + \frac{1}{4} =$

3. $\frac{3}{8} + \frac{1}{3} =$

4. $\frac{1}{4} + \frac{2}{7} =$

5. $\frac{1}{4} + \frac{2}{5} =$

$$\frac{1}{4} = \frac{5}{20}$$
$$+\frac{2}{5} = \frac{8}{20}$$
$$\frac{13}{20}$$

6. $\frac{2}{3} + \frac{1}{6} =$

7. $\frac{2}{9} + \frac{1}{3} =$

8. $\frac{1}{5}$
$+\frac{1}{3}$

9. $\frac{3}{4}$
$+\frac{1}{6}$

10. $\frac{1}{7}$
$+\frac{1}{2}$

11. $\frac{5}{8}$
$+\frac{1}{5}$

12. $\frac{3}{10}$
$+\frac{1}{3}$

Adding Fractions with Different Denominators

Compare the denominators before adding. Find a common denominator and add only the numerators. If the answer is an improper fraction, change to a whole or mixed number. Reduce if possible.

Use These Steps

Add $\frac{3}{4} + \frac{1}{3}$

1. Multiply the denominators to find the common denominator.

   $\frac{3}{4} = \frac{}{12}$
   $+\frac{1}{3} = \frac{}{12}$

2. Raise each fraction to higher terms.

   $\frac{3}{4} = \frac{9}{12}$
   $+\frac{1}{3} = \frac{4}{12}$

3. Add. The sum is an improper fraction. Change to a mixed number.

   $\frac{3}{4} = \frac{9}{12}$
   $+\frac{1}{3} = \frac{4}{12}$
   $\frac{13}{12} = 1\frac{1}{12}$

Add. Change improper fractions to whole or mixed numbers. Reduce if possible.

1. $\frac{7}{8} + \frac{2}{3} =$

   $\frac{7}{8} = \frac{21}{24}$
   $+\frac{2}{3} = \frac{16}{24}$
   $\frac{37}{24} = 1\frac{13}{24}$

2. $\frac{4}{5} + \frac{3}{4} =$

3. $\frac{1}{3} + \frac{9}{10} =$

4. $\frac{5}{6} + \frac{2}{7} =$

5. $\frac{1}{2} + \frac{3}{4} + \frac{1}{5} =$

   $\frac{1}{2} = \frac{10}{20}$
   $\frac{3}{4} = \frac{15}{20}$
   $+\frac{1}{5} = \frac{4}{20}$
   $\frac{29}{20} = 1\frac{9}{20}$

6. $\frac{7}{8} + \frac{3}{4} + \frac{2}{3} =$

7. $\frac{2}{7} + \frac{3}{5} + \frac{1}{4} =$

8. Wilfred bought $\frac{5}{6}$ pound of corned beef, $\frac{3}{4}$ pound of turkey, and $\frac{1}{2}$ pound of cheese for sandwiches. How much did the three purchases weigh in all?

9. For dessert, Wilfred bought $\frac{2}{3}$ pound of apples, $\frac{3}{4}$ pound of pears, and $\frac{1}{2}$ pound of raisins. How much did the three purchases weigh in all?

Answer _____

Answer _____

Finding the Lowest Common Denominator

Another way to find a common denominator is by making a list of multiples of each denominator. A multiple of a number is the number multiplied by 1, 2, 3, and so on. For example, some multiples of 2 are 2, 4, 6, 8 and 10.

After listing multiples for each denominator, find the smallest number that appears on both lists. This number is the lowest common denominator (LCD) for the fractions.

Use These Steps

Add $\frac{1}{6} + \frac{3}{4}$

1. List multiples of each denominator. The smallest number on both lists is 12. 12 is the LCD.

   $\frac{1}{6}$ 6 (12) 18 24

   $\frac{3}{4}$ 4 8 (12) 16

2. Raise each fraction to higher terms with 12 as the LCD.

   $\frac{1}{6} = \frac{2}{12}$

   $+\frac{3}{4} = \frac{9}{12}$

3. Add.

   $\frac{2}{12}$

   $+\frac{9}{12}$

   $\overline{\frac{11}{12}}$

List the multiples of each denominator to find the LCD. Change the fractions. Then add. Reduce if possible.

1. $\frac{1}{6} + \frac{1}{9} =$

   $\frac{1}{6}$ 6 12 (18) 24

   $\frac{1}{9}$ 9 (18) 27

   $\frac{1}{6} = \frac{3}{18}$

   $+\frac{1}{9} = \frac{2}{18}$

   $\overline{\frac{5}{18}}$

2. $\frac{5}{6} + \frac{1}{4} =$

3. $\frac{5}{8} + \frac{3}{10} =$

4. $\frac{1}{12} + \frac{3}{16} =$

5. $\frac{1}{6}$

   $+\frac{7}{10}$

6. $\frac{5}{12}$

   $+\frac{17}{18}$

7. $\frac{4}{9}$

   $+\frac{7}{12}$

8. $\frac{3}{8}$

   $+\frac{5}{6}$

Mixed Review

Write each set of fractions with common denominators.

1. $\frac{1}{2}$ and $\frac{1}{4}$
2. $\frac{1}{3}$ and $\frac{2}{5}$
3. $\frac{6}{7}$ and $\frac{2}{21}$
4. $\frac{1}{6}$ and $\frac{5}{8}$
5. $\frac{5}{9}$ and $\frac{3}{10}$

Find a common denominator. Add. Change improper fractions to whole or mixed numbers. Reduce if possible.

6. $\frac{1}{6} + \frac{1}{3} =$
7. $\frac{3}{8} + \frac{1}{4} =$
8. $\frac{2}{9} + \frac{2}{3} =$
9. $\frac{7}{16} + \frac{3}{4} =$

10. $\frac{2}{3} + \frac{3}{4} =$
11. $\frac{5}{6} + \frac{3}{5} =$
12. $\frac{3}{5} + \frac{1}{4} =$
13. $\frac{1}{2} + \frac{7}{9} =$

14. $\frac{7}{10} + \frac{4}{5}$
15. $\frac{5}{8} + \frac{1}{2}$
16. $\frac{2}{7} + \frac{1}{4}$
17. $\frac{1}{5} + \frac{1}{3}$
18. $\frac{3}{4} + \frac{2}{3}$

19. $\frac{4}{5} + \frac{3}{10} + \frac{1}{2} =$
20. $\frac{3}{4} + \frac{3}{8} + \frac{3}{16} =$
21. $\frac{5}{6} + \frac{1}{3} + \frac{2}{9} =$

22. $\frac{11}{12} + \frac{1}{4} + \frac{2}{3}$
23. $\frac{4}{15} + \frac{2}{5} + \frac{1}{2}$
24. $\frac{3}{7} + \frac{2}{21} + \frac{6}{14}$
25. $\frac{1}{2} + \frac{3}{4} + \frac{5}{6}$
26. $\frac{8}{9} + \frac{3}{18} + \frac{1}{3}$

Adding Mixed Numbers

When you add mixed numbers, you may need to find a common denominator.

Use These Steps

Add $5\frac{1}{3} + 2\frac{1}{8}$

1. Find a common denominator for the fractions.

$$5\frac{1}{3} = 5\frac{}{24}$$
$$+2\frac{1}{8} = 2\frac{}{24}$$

2. Raise each fraction to higher terms with the new denominator.

$$5\frac{1}{3} = 5\frac{8}{24}$$
$$+2\frac{1}{8} = 2\frac{3}{24}$$

3. Add the fractions. Add the whole numbers.

$$5\frac{1}{3} = 5\frac{8}{24}$$
$$+2\frac{1}{8} = 2\frac{3}{24}$$
$$7\frac{11}{24}$$

Add. Reduce if possible.

1. $2\frac{1}{4} + 3\frac{1}{8} =$

    $2\frac{1}{4} = 2\frac{2}{8}$
    $+3\frac{1}{8} = 3\frac{1}{8}$
    $\phantom{+3\frac{1}{8} =} 5\frac{3}{8}$

2. $3\frac{2}{3} + 4\frac{1}{9} =$

3. $6\frac{4}{7} + 5\frac{9}{14} =$

4. $7\frac{1}{2} + 12\frac{5}{6} =$

5. $3\frac{2}{3} + 4\frac{1}{4} =$

6. $1\frac{5}{8} + 4\frac{1}{6} =$

7. $8\frac{3}{5} + 9\frac{2}{7} =$

8. $10\frac{3}{10} + 4\frac{1}{3} =$

9. $7\frac{1}{6} + 2\frac{2}{3} + 4\frac{7}{9} =$

    $7\frac{1}{6} = 7\frac{3}{18}$
    $2\frac{2}{3} = 2\frac{12}{18}$
    $+4\frac{7}{9} = 4\frac{14}{18}$
    $\phantom{+4\frac{7}{9} =} 13\frac{29}{18} = 14\frac{11}{18}$

10. $1\frac{1}{2} + 6\frac{3}{4} + 2\frac{5}{6} =$

11. $9\frac{1}{5} + 6\frac{7}{10} + 3\frac{1}{2} =$

Adding Mixed Numbers, Whole Numbers, and Fractions

When adding mixed numbers, whole numbers, and fractions, compare the denominators of the fractions. Then find a common denominator and raise fractions to higher terms.

Use These Steps

Add $2\frac{1}{2} + \frac{5}{6} + 3$

1. Compare the denominators. 2 divides evenly into 6. Raise $\frac{1}{2}$ to higher terms with 6 as the denominator.

$$2\frac{1}{2} = 2\frac{3}{6}$$
$$\frac{5}{6} = \frac{5}{6}$$
$$+3 = 3$$

2. Add the fractions. Add the whole numbers.

$$2\frac{1}{2} = 2\frac{3}{6}$$
$$\frac{5}{6} = \frac{5}{6}$$
$$+3 = 3$$
$$\overline{5\frac{8}{6}}$$

3. Change the improper fraction in the sum to a mixed number. Add the new mixed number to the whole number. Reduce.

$$5\frac{8}{6} = 5 + 1\frac{2}{6} = 6\frac{2}{6} = 6\frac{1}{3}$$

Add. Reduce if possible.

1. $3\frac{7}{8} + \frac{1}{3} + 6 =$

    $3\frac{7}{8} = 3\frac{21}{24}$
    $\frac{1}{3} = \frac{8}{24}$
    $+6 = 6$
    $\overline{9\frac{29}{24} = 9 + 1\frac{5}{24} = 10\frac{5}{24}}$

2. $4\frac{1}{2} + 7 =$

3. $8\frac{3}{4} + \frac{5}{6} =$

4. $7 + 2\frac{1}{8} + \frac{3}{7} =$

5. $\frac{9}{10} + 3\frac{1}{5} + 20 =$

6. $\frac{5}{12} + 2\frac{1}{6} + 4 =$

7.
$$3$$
$$\frac{5}{6}$$
$$+2\frac{1}{2}$$

8.
$$15\frac{1}{2}$$
$$7$$
$$+2\frac{3}{12}$$

9.
$$8\frac{9}{10}$$
$$6$$
$$+2\frac{8}{15}$$

10.
$$\frac{1}{4}$$
$$\frac{1}{6}$$
$$+9\frac{1}{12}$$

11.
$$18$$
$$24\frac{2}{5}$$
$$+12\frac{7}{30}$$

Problem Solving: Using Time Records

Several parents serve as teachers' aides in the Southside Elementary School. Reiko writes down the hours worked each day. At the end of the week, she adds the hours to get the total hours each person worked.

Aide	November 5					Hours Worked
	M	Tu	W	Th	F	Total
Tim	$2\frac{2}{3}$	$2\frac{1}{3}$	3	$4\frac{3}{4}$	$3\frac{1}{4}$	16
Jorge	$3\frac{3}{4}$	$5\frac{1}{2}$	$6\frac{1}{3}$	$3\frac{3}{4}$	$5\frac{2}{3}$	
Sue	$6\frac{1}{3}$	7	$6\frac{1}{2}$	$4\frac{1}{2}$	$5\frac{2}{3}$	
Rachel	$6\frac{2}{3}$	$5\frac{2}{3}$	$5\frac{1}{2}$	$5\frac{1}{2}$	$5\frac{2}{3}$	

Example Find the total hours Tim worked the week of November 5. Write the amount in the chart.

▶ **Step 1.** Multiply 3×4 to find the common denominator. Change all fractions to higher terms with 12 as the denominator. Add.

$2\frac{2}{3} = 2\frac{8}{12}$
$2\frac{1}{3} = 2\frac{4}{12}$
$3 \quad = 3$
$4\frac{3}{4} = 4\frac{9}{12}$
$+ 3\frac{1}{4} = 3\frac{3}{12}$
$\overline{14\frac{24}{12}}$

▶ **Step 2.** Change $\frac{24}{12}$ to 2. Add 2 to 14.

$$14 + 2 = 16$$

The week of November 5, Tim worked a total of 16 hours. 16 is written in the total column at the end of Tim's row.

Solve. Reduce if possible.

1. Find the total hours that Jorge worked. Write the amount in the chart.

2. Find the total hours for Sue. Write the total in the chart.

   Answer_____   Answer_____

3. Find the total hours Rachel worked. Write this total in the chart.

4. Find the grand total of the hours all four teachers' aides worked.

   Answer_____   Answer_____

Example During the week of September 24, the teachers' aides worked a grand total of 48 hours. Ray worked 12 hours that week. What fraction of the grand total did Ray work during the week of September 24?

Aide	September			
	3	10	17	24
Ray	0	$8\frac{1}{2}$	17	12
June	5	$5\frac{1}{2}$	15	6
Carlos	3	2	$4\frac{1}{2}$	$12\frac{3}{4}$
Kenji	7	8	$13\frac{1}{2}$	$17\frac{1}{4}$
Total	15	24	50	48

Step 1. Set up a fraction with the number of hours Ray worked, 12, as the numerator. The grand total, 48, is the denominator.

$$\frac{12}{48}$$

Step 2. Reduce the fraction to lowest terms.

$$\frac{12}{48} = \frac{1}{4}$$

Ray worked $\frac{1}{4}$ of the grand total during the week of September 24.

Solve. Reduce if possible.

5. What fraction of the grand total for the week of September 17 did June work?

    Answer _____

6. What fraction of the grand total for the week of September 10 did Kenji work?

    Answer _____

7. What fraction of the grand total for the week of September 3 did Carlos work?

    Answer _____

8. What fraction of the grand total for the week of September 10 did Carlos work?

    Answer _____

Unit 2 Review

Add. Change improper fractions to mixed or whole numbers. Reduce if possible.

1. $\frac{1}{3} + \frac{1}{3} =$
2. $\frac{1}{4} + \frac{2}{4} =$
3. $\frac{3}{8} + \frac{2}{8} =$
4. $\frac{12}{15} + \frac{2}{15} =$

5. $\frac{1}{16} + \frac{12}{16}$
6. $\frac{7}{10} + \frac{2}{10}$
7. $\frac{4}{25} + \frac{13}{25}$
8. $\frac{3}{50} + \frac{26}{50}$
9. $\frac{14}{100} + \frac{27}{100}$

10. $\frac{2}{6} + \frac{1}{6} =$
11. $\frac{5}{9} + \frac{5}{9} =$
12. $\frac{7}{11} + \frac{5}{11} =$
13. $\frac{4}{5} + \frac{3}{5} + \frac{4}{5} =$

14. $\frac{2}{7} + \frac{5}{7}$
15. $\frac{5}{8} + \frac{7}{8}$
16. $\frac{3}{6} + \frac{4}{6}$
17. $\frac{3}{10} + \frac{3}{10}$
18. $\frac{5}{12} + \frac{7}{12}$

19. $3\frac{1}{2} + 2\frac{1}{2} =$
20. $16\frac{3}{8} + 10\frac{7}{8} =$
21. $21\frac{2}{9} + 29\frac{4}{9} =$
22. $1\frac{9}{10} + 2 + \frac{7}{10} =$

23. $5\frac{1}{4} + 6\frac{1}{4}$
24. $10\frac{7}{8} + 12\frac{5}{8}$
25. $9\frac{7}{10} + 15\frac{1}{10}$
26. $18\frac{4}{18} + 3\frac{16}{18}$
27. $7\frac{3}{20} + 4\frac{9}{20}$

28. $\frac{2}{5} + \frac{3}{10} =$
29. $\frac{3}{16} + \frac{1}{2} =$
30. $\frac{5}{18} + \frac{3}{9} =$
31. $\frac{3}{4} + \frac{1}{8} =$

32. $\frac{2}{7} + \frac{1}{14}$
33. $\frac{1}{2} + \frac{5}{12}$
34. $\frac{1}{5} + \frac{1}{10}$
35. $\frac{4}{15} + \frac{2}{3}$
36. $\frac{19}{25} + \frac{1}{5}$

Add. Change improper fractions to mixed or whole numbers. Reduce if possible.

37. $\dfrac{1}{3} + \dfrac{1}{6} =$

38. $\dfrac{1}{2} + \dfrac{5}{10} =$

39. $\dfrac{3}{12} + \dfrac{5}{6} =$

40. $\dfrac{2}{3} + \dfrac{5}{6} + \dfrac{1}{2} =$

41. $\dfrac{4}{16}$
$+\dfrac{3}{8}$

42. $\dfrac{2}{3}$
$+\dfrac{5}{12}$

43. $\dfrac{2}{7}$
$+\dfrac{15}{21}$

44. $\dfrac{7}{30}$
$+\dfrac{14}{15}$

45. $\dfrac{1}{2}$
$+\dfrac{9}{10}$

46. $\dfrac{2}{6} + \dfrac{1}{4} =$

47. $\dfrac{5}{8} + \dfrac{2}{3} =$

48. $\dfrac{3}{4} + \dfrac{1}{7} =$

49. $\dfrac{3}{10} + \dfrac{2}{3} + \dfrac{1}{2} =$

50. $\dfrac{1}{2}$
$+\dfrac{2}{3}$

51. $\dfrac{3}{4}$
$+\dfrac{5}{6}$

52. $\dfrac{2}{5}$
$+\dfrac{1}{3}$

53. $\dfrac{5}{9}$
$+\dfrac{1}{2}$

54. $\dfrac{5}{6}$
$+\dfrac{3}{8}$

55. $3\dfrac{1}{2} + 6\dfrac{2}{3} =$

56. $3\dfrac{3}{5} + 9\dfrac{1}{4} + 2 =$

57. $\dfrac{1}{8} + 14\dfrac{2}{6} + 3 =$

58. $5\dfrac{1}{4}$
$+6\dfrac{1}{3}$

59. $1\dfrac{2}{5}$
$+9\dfrac{3}{4}$

60. $8\dfrac{9}{10}$
$+6\dfrac{5}{6}$

61. $15\dfrac{1}{9}$
$+\;2\dfrac{1}{2}$

62. $27\dfrac{5}{6}$
$+19\dfrac{1}{8}$

Below is a list of the problems in this review and the pages on which the skills are taught. If you missed any problems, turn to the pages listed and practice the skills. Then correct the problems you missed in the Unit Review.

Problems	Pages
1-9	41-42
10-18	44-46
19-27	49-51
28-36	55-56

Problems	Pages
37-45	57-59
46-54	62-65
55-62	67-68

Unit 3 SUBTRACTING FRACTIONS

Subtraction problems with fractions usually ask you to find the difference between two numbers or how much is left after using part of something. You may have needed to subtract fractions when working with craft projects or figuring distance on a car trip.

In this unit you will learn how to subtract fractions and mixed numbers, how to subtract fractions from whole numbers, and how to solve word problems using fractions.

Getting Ready

You should be familiar with the skills on this page and the next before you begin this unit. To check your answers, turn to page 182.

 When subtracting, you may need to borrow.

Subtract.

1. 15
    − 7
    ———
    8

2. 92
    − 39

3. 40
    − 18

4. 700
    − 26

5. 401
    − 193

6. 831
    − 256

7. 284
    − 194

8. 3,106
    − 928

9. 8,000
    − 6,241

10. 26,002
    − 9,607

For review, see pages 66–78 in Math Matters for Adults, Whole Numbers.

Getting Ready

 To raise a fraction to higher terms with a given denominator, multiply the numerator and denominator by the same number.

Write each fraction using the given denominator.

11. $\frac{3}{4} = \frac{6}{8}$
12. $\frac{1}{2} = \frac{}{10}$
13. $\frac{2}{3} = \frac{}{12}$
14. $\frac{3}{5} = \frac{}{15}$
15. $\frac{3}{7} = \frac{}{21}$

16. $\frac{4}{9} = \frac{}{27}$
17. $\frac{1}{10} = \frac{}{30}$
18. $\frac{5}{8} = \frac{}{40}$
19. $\frac{7}{12} = \frac{}{48}$
20. $\frac{5}{15} = \frac{}{60}$

For review, see Unit 1, pages 19–20.

 Before adding or subtracting fractions, you may need to find a common denominator.

Write each set of fractions with a common denominator.

21. $\frac{1}{4}$ and $\frac{1}{2}$
    $\frac{1}{4}$ and $\frac{2}{4}$
22. $\frac{2}{3}$ and $\frac{3}{4}$
23. $\frac{5}{7}$ and $\frac{3}{5}$
24. $3\frac{3}{12}$ and $1\frac{1}{4}$
25. $2\frac{1}{10}$ and $6\frac{3}{5}$

26. $\frac{3}{8}$ and $\frac{5}{6}$
27. $2\frac{4}{9}$ and $5\frac{1}{18}$
28. $\frac{3}{10}$ and $\frac{3}{25}$
29. $\frac{2}{3}$ and $\frac{2}{9}$
30. $\frac{1}{6}$ and $\frac{1}{4}$

For review, see Unit 1, page 25.

 To reduce fractions to lowest terms, divide the numerator and the denominator by the same number.

Reduce each fraction to lowest terms. If the fraction is in lowest terms, write LT.

31. $\frac{3}{6} = \frac{1}{2}$
32. $\frac{10}{11} =$
33. $\frac{8}{12} =$
34. $2\frac{15}{25} =$
35. $5\frac{7}{15} =$

36. $\frac{2}{9} =$
37. $\frac{4}{8} =$
38. $\frac{4}{16} =$
39. $2\frac{7}{21} =$
40. $\frac{10}{35} =$

For review, see Unit 1, pages 15–16.

Subtracting Fractions with the Same Denominator

To subtract fractions with the same denominator, subtract only the numerators. Keep the denominator.

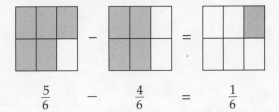

$$\frac{5}{6} - \frac{4}{6} = \frac{1}{6}$$

Use These Steps

Subtract $\frac{5}{8} - \frac{2}{8}$

1. The denominators are the same. Write the denominator under the fraction bar.

   $$\frac{5}{8} - \frac{2}{8} = \frac{}{8}$$

2. Subtract only the numerators. $5 - 2 = 3$. Write the difference over the denominator.

   $$\frac{5}{8} - \frac{2}{8} = \frac{3}{8}$$

Shade in the figures to show the fractions. Subtract the fractions. Then shade in the final figure to show the difference.

1.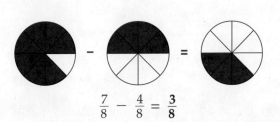

   $$\frac{7}{8} - \frac{4}{8} = \frac{3}{8}$$

2.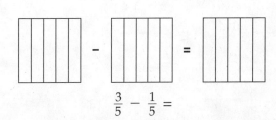

   $$\frac{3}{5} - \frac{1}{5} =$$

3.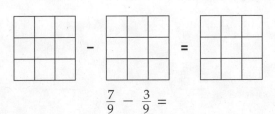

   $$\frac{7}{9} - \frac{3}{9} =$$

4.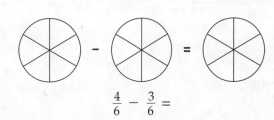

   $$\frac{4}{6} - \frac{3}{6} =$$

Subtract.

5. $\frac{5}{9} - \frac{3}{9} = \frac{2}{9}$

6. $\frac{4}{6} - \frac{3}{6} =$

7. $\frac{4}{7} - \frac{1}{7} =$

8. $\frac{4}{5} - \frac{2}{5} =$

Subtracting Fractions with the Same Denominator

Fraction problems may be set up horizontally or vertically. Be sure the denominators are the same and subtract only the numerators.

Use These Steps

Subtract $\frac{7}{10} - \frac{4}{10}$

1. The denominators are the same. Write the denominator under the fraction bar.

$$\frac{7}{10} - \frac{4}{10} = \frac{}{10}$$

2. Subtract only the numerators. $7 - 4 = 3$. Write the difference over the denominator.

$$\frac{7}{10} - \frac{4}{10} = \frac{3}{10}$$

Subtract.

1. $\frac{7}{14} - \frac{4}{14} = \frac{3}{14}$

2. $\frac{9}{16} - \frac{2}{16}$

3. $\frac{15}{25} - \frac{9}{25}$

4. $\frac{11}{12} - \frac{4}{12}$

5. $\frac{5}{7} - \frac{3}{7}$

6. $\frac{6}{9} - \frac{2}{9} =$

7. $\frac{8}{10} - \frac{1}{10} =$

8. $\frac{13}{15} - \frac{5}{15} =$

9. $\frac{17}{20} - \frac{10}{20} =$

10. $\frac{29}{30} - \frac{18}{30}$

11. $\frac{11}{50} - \frac{8}{50}$

12. $\frac{31}{42} - \frac{12}{42}$

13. $\frac{9}{11} - \frac{6}{11}$

14. $\frac{99}{100} - \frac{68}{100}$

15. $\frac{60}{75} - \frac{37}{75} =$

16. $\frac{50}{63} - \frac{25}{63} =$

17. $\frac{18}{35} - \frac{9}{35} =$

18. $\frac{12}{18} - \frac{7}{18} =$

19. Louise had $\frac{3}{4}$ pound of candy. She gave $\frac{2}{4}$ pound to her friend. How much candy is left?

20. Roland had $\frac{7}{8}$ yard of plastic tubing. He used $\frac{4}{8}$ yard. How much plastic tubing was left?

Answer_____

Answer_____

Subtracting and Reducing Fractions

When subtracting fractions, you may get an answer that can be reduced. Always reduce answers to lowest terms.

Use These Steps

Subtract $\frac{3}{16} - \frac{1}{16}$

1. The denominators are the same. Write the denominator under the fraction bar.

   $\frac{3}{16} - \frac{1}{16} = \frac{}{16}$

2. Subtract the numerators. Write the difference over the denominator.

   $\frac{3}{16} - \frac{1}{16} = \frac{2}{16}$

3. Reduce the answer to lowest terms.

   $\frac{2}{16} = \frac{2 \div 2}{16 \div 2} = \frac{1}{8}$

Subtract. Reduce to lowest terms.

1. $\frac{4}{9} - \frac{1}{9} =$
   $\frac{3}{9} = \frac{3 \div 3}{9 \div 3} = \frac{1}{3}$

2. $\frac{10}{12} - \frac{1}{12} =$

3. $\frac{5}{8} - \frac{3}{8} =$

4. $\frac{7}{10} - \frac{2}{10} =$

5. $\frac{3}{4} - \frac{1}{4} =$

6. $\frac{5}{6} - \frac{3}{6} =$

7. $\frac{13}{14} - \frac{7}{14} =$

8. $\frac{12}{15} - \frac{2}{15} =$

9. $\frac{4}{6} - \frac{1}{6}$

10. $\frac{7}{8} - \frac{1}{8}$

11. $\frac{5}{10} - \frac{3}{10}$

12. $\frac{17}{20} - \frac{2}{20}$

13. $\frac{20}{21} - \frac{6}{21}$

14. $\frac{20}{27} - \frac{2}{27}$

15. $\frac{24}{30} - \frac{19}{30}$

16. $\frac{34}{35} - \frac{20}{35}$

17. $\frac{46}{50} - \frac{21}{50}$

18. $\frac{90}{100} - \frac{15}{100}$

19. Jim bought $\frac{3}{4}$ pound of fresh fish. He used $\frac{1}{4}$ pound to make lunch. How much did he have left?

20. Anne bought $\frac{9}{16}$ pound of peaches. She used $\frac{7}{16}$ pound to make jam. How much did she have left?

Answer _____

Answer _____

Subtracting Mixed Numbers

You know that a mixed number has a whole number and a fraction. When subtracting mixed numbers, subtract the fractions and then the whole numbers. Reduce the answer if possible.

Use These Steps

Subtract $3\frac{5}{9} - 1\frac{2}{9}$

1. Line up the fractions and the whole numbers in columns.

$$3\frac{5}{9}$$
$$-1\frac{2}{9}$$

2. Subtract the fractions.

$$3\frac{5}{9}$$
$$-1\frac{2}{9}$$
$$\frac{3}{9}$$

3. Subtract the whole numbers. Reduce the fraction to lowest terms.

$$3\frac{5}{9}$$
$$-1\frac{2}{9}$$
$$2\frac{3}{9} = 2\frac{1}{3}$$

Subtract. Reduce if possible.

1. $5\frac{5}{6} - 1\frac{2}{6} =$

$$5\frac{5}{6}$$
$$-1\frac{2}{6}$$
$$4\frac{3}{6} = 4\frac{1}{2}$$

2. $10\frac{7}{8} - 5\frac{5}{8} =$

3. $7\frac{3}{4} - 1\frac{1}{4} =$

4. $15\frac{11}{12} - 8\frac{8}{12} =$

5. $18\frac{17}{20}$
   $-15\frac{2}{20}$

6. $13\frac{23}{28}$
   $-2\frac{19}{28}$

7. $26\frac{27}{30}$
   $-19\frac{12}{30}$

8. $15\frac{30}{45}$
   $-11\frac{21}{45}$

9. $31\frac{20}{50}$
   $-27\frac{15}{50}$

10. $7\frac{9}{10} - 4\frac{3}{10} =$

11. $13\frac{7}{16} - 2\frac{4}{16} =$

12. $15\frac{11}{15} - 9\frac{8}{15} =$

13. $20\frac{15}{18} - 6\frac{3}{18} =$

14. Willie bought $2\frac{7}{8}$ yards of material to make seat covers for his truck. He used $1\frac{3}{8}$ yards for one seat cover. How many yards does he have left?

    Answer _____

15. Willie also bought $12\frac{3}{4}$ yards of trim. He used $6\frac{1}{4}$ yards for one seat cover. How much trim does he have left?

    Answer _____

Real-Life Application — On the Job

Adrienne works at the city animal shelter. Her job is to feed the dogs and cats twice a day. She measures the animals' food from large bags.

Example Adrienne used $12\frac{3}{8}$ pounds of dog food from a bag that had $21\frac{7}{8}$ pounds in it. How much was left in the bag?
To find the amount left, subtract. Reduce to lowest terms.

$$21\frac{7}{8}$$
$$-12\frac{3}{8}$$
$$\overline{9\frac{4}{8}} = 9\frac{1}{2}$$

There were $9\frac{1}{2}$ pounds of dog food left in the bag.

Solve. Reduce if possible.

1. Adrienne needed $12\frac{7}{8}$ pounds of puppy food to feed the puppies. The open bag of food only had $10\frac{3}{8}$ pounds in it. How much more puppy food did she need?

   Answer_____

2. It takes $10\frac{5}{16}$ pounds of cat food to feed all the cats in the shelter. Adrienne gets this amount from a bag that has $14\frac{9}{16}$ pounds in it. How much is left in the bag after she feeds all the cats?

   Answer_____

3. Adrienne gives each of the animals water from a container that holds $15\frac{2}{3}$ cups. She uses $10\frac{1}{3}$ cups. How many cups of water are left in the container?

   Answer_____

4. Adrienne checks the amount of food left at the end of each day. Today she counted $50\frac{3}{4}$ pounds of dog food and $28\frac{1}{4}$ pounds of cat food. How much more dog food than cat food is left?

   Answer_____

Subtracting Fractions from Mixed Numbers

When you subtract a fraction from a mixed number, first subtract the fractions. Since there is no whole number to subtract, bring down the whole number part of the mixed number into the answer.

Use These Steps

Subtract $5\frac{4}{10} - \frac{2}{10}$

1. Line up the fractions.

$$5\frac{4}{10}$$
$$-\frac{2}{10}$$

2. Subtract the fractions. Bring down the whole number.

$$5\frac{4}{10}$$
$$-\frac{2}{10}$$
$$5\frac{2}{10}$$

3. Reduce the fraction to lowest terms.

$$5\frac{2}{10} = 5\frac{1}{5}$$

Subtract. Reduce if possible.

1. $3\frac{5}{16} - \frac{3}{16} =$

$$3\frac{5}{16}$$
$$-\frac{3}{16}$$
$$3\frac{2}{16} = 3\frac{1}{8}$$

2. $9\frac{7}{12} - \frac{1}{12} =$

3. $11\frac{2}{3} - \frac{1}{3} =$

4. $4\frac{9}{15} - \frac{4}{15} =$

5. $36\frac{18}{25} - \frac{3}{25}$

6. $42\frac{12}{21} - \frac{5}{21}$

7. $50\frac{26}{32} - \frac{16}{32}$

8. $61\frac{48}{49} - \frac{34}{49}$

9. $79\frac{33}{100} - \frac{24}{100}$

10. $12\frac{9}{10} - \frac{3}{10} =$

11. $20\frac{5}{18} - \frac{2}{18} =$

12. $14\frac{11}{20} - \frac{1}{20} =$

13. $22\frac{9}{17} - \frac{2}{17} =$

14. $59\frac{14}{15} - \frac{12}{15}$

15. $32\frac{9}{16} - \frac{5}{16}$

16. $44\frac{20}{27} - \frac{2}{27}$

17. $95\frac{27}{50} - \frac{2}{50}$

18. $102\frac{60}{75} - \frac{15}{75}$

Subtracting Whole Numbers from Mixed Numbers

When you subtract a whole number from a mixed number, first bring down the fraction. Then subtract the whole numbers.

Use These Steps

Subtract $9\frac{7}{8} - 3$

1. Line up the whole numbers.

$$9\frac{7}{8} \\ -\,3 \\ \hline$$

2. Bring down the fraction.

$$9\frac{7}{8} \\ -\,3 \\ \hline \frac{7}{8}$$

3. Subtract the whole numbers.

$$9\frac{7}{8} \\ -\,3 \\ \hline 6\frac{7}{8}$$

Subtract.

1. $12\frac{5}{11} - 9 =$

$$12\frac{5}{11} \\ -\,9 \\ \hline 3\frac{5}{11}$$

2. $15\frac{2}{3} - 7 =$

3. $19\frac{1}{7} - 13 =$

4. $5\frac{2}{9} - 3 =$

5. $57\frac{9}{16} \\ -\,32 \\ \hline$

6. $84\frac{22}{35} \\ -\,19 \\ \hline$

7. $99\frac{9}{10} \\ -\,80 \\ \hline$

8. $117\frac{27}{50} \\ -\,94 \\ \hline$

9. $125\frac{73}{100} \\ -\,46 \\ \hline$

10. $23\frac{7}{13} - 14 =$

11. $30\frac{10}{21} - 15 =$

12. $25\frac{9}{20} - 18 =$

13. $41\frac{12}{25} - 37 =$

14. Joseph had $3\frac{1}{2}$ pounds of peanuts. He used 2 pounds to feed the squirrels. How much did he have left?

Answer _____

15. Joseph had $1\frac{3}{4}$ gallons of paint. He used 1 gallon to paint the kitchen. How much paint did he have left?

Answer _____

Mixed Review

Reduce each fraction to lowest terms.

1. $\dfrac{12}{16} =$
2. $\dfrac{2}{10} =$
3. $\dfrac{6}{9} =$
4. $\dfrac{2}{4} =$
5. $\dfrac{16}{18} =$

Add or subtract. Reduce if possible.

6. $\dfrac{3}{5} + \dfrac{1}{5} =$
7. $3\dfrac{6}{9} - \dfrac{3}{9} =$
8. $\dfrac{5}{6} - \dfrac{2}{6} =$
9. $1\dfrac{5}{8} + 5\dfrac{1}{8} =$

10. $\dfrac{1}{7} + \dfrac{3}{7} =$
11. $2\dfrac{3}{10} + 4\dfrac{3}{10} =$
12. $5\dfrac{7}{12} - 1\dfrac{1}{12} =$
13. $\dfrac{8}{15} - \dfrac{3}{15} =$

14. $14\dfrac{3}{20} + 18\dfrac{2}{20}$
15. $1\dfrac{8}{9} - \dfrac{2}{9}$
16. $\dfrac{12}{25} - \dfrac{7}{25}$
17. $1\dfrac{3}{13} + 9\dfrac{7}{13}$
18. $\dfrac{9}{14} - \dfrac{5}{14}$

19. $\dfrac{11}{35} + \dfrac{16}{35}$
20. $2\dfrac{16}{27} - 2\dfrac{13}{27}$
21. $5\dfrac{4}{30} + 6\dfrac{16}{30}$
22. $10\dfrac{8}{21} - 1\dfrac{5}{21}$
23. $\dfrac{15}{42} - \dfrac{9}{42}$

24. $2\dfrac{9}{10} + 3 =$
25. $\dfrac{7}{16} - \dfrac{3}{16} =$
26. $10\dfrac{7}{18} + 3\dfrac{2}{18} =$
27. $\dfrac{6}{7} - \dfrac{3}{7} =$

28. $21\dfrac{3}{5} - 5$
29. $4\dfrac{19}{49} + \dfrac{22}{49}$
30. $\dfrac{63}{75} - \dfrac{13}{75}$
31. $12\dfrac{27}{64} - 4\dfrac{19}{64}$
32. $43\dfrac{98}{100} + 7\dfrac{1}{100}$

33. $19\dfrac{2}{3} - 11 =$
34. $\dfrac{6}{20} + \dfrac{3}{20} =$
35. $\dfrac{7}{10} - \dfrac{5}{10} =$
36. $6\dfrac{5}{12} - 3\dfrac{1}{12} =$

Problem Solving: **Using a Ruler**

The ruler below is divided into inches and fractions of an inch: halves, fourths, eighths, and sixteenths.

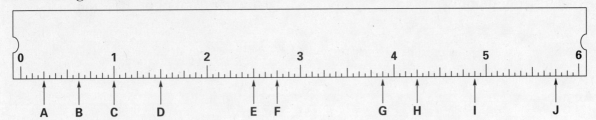

To find the distance between two points on the ruler, first find the distance from zero for both points. Then subtract.

Example How far is point F from point A?

▶ **Step 1.** Find point F. The nearest whole inch mark to the left is 2. The mark at the arrow is $\frac{3}{4}$ inch from 2. Point F is $2\frac{3}{4}$ inches from zero.

▶ **Step 2.** Find point A. The nearest whole inch mark to the left is zero. The mark at the arrow is $\frac{1}{4}$ inch from zero. Point A is $\frac{1}{4}$ inch from zero.

▶ **Step 3.** Subtract. Reduce to lowest terms.

$$2\frac{3}{4} - \frac{1}{4} = 2\frac{2}{4} = 2\frac{1}{2}$$

Point F is $2\frac{1}{2}$ inches from point A.

Use the ruler to answer the questions. Reduce if possible.

1. How far from point B is point I?

   Answer _____

2. How far from point A is point J?

   Answer _____

3. How far from point B is point G?

   Answer _____

4. How far from point C is point D?

   Answer _____

Use the ruler to answer the questions. Reduce if possible.

5. How far from point D is point E?

   Answer_____

6. How far from point F is point J?

   Answer_____

7. How far from point C is point F?

   Answer_____

8. How far from point G is point I?

   Answer_____

9. Mike drew a line from point H to point J on the ruler. How long was the line he drew?

   Answer_____

10. Tony drew a line from point A to point H on the ruler. How long was the line he drew?

    Answer_____

Solve. Reduce if possible.

11. Sam is making a picture frame that is $3\frac{1}{2}$ inches wide. He has a piece of picture frame molding that is $4\frac{1}{2}$ inches long. If he cuts off $3\frac{1}{2}$ inches, how much molding will he have left?

    Answer_____

12. Sam's picture frame is $6\frac{1}{4}$ inches long. He has another piece of molding that is $9\frac{3}{4}$ inches long. If he cuts off $6\frac{1}{4}$ inches, how much will he have left?

    Answer_____

13. Helen measures how tall her children are each year. Last year Helen's son was $34\frac{1}{4}$ inches tall. Now he is $36\frac{3}{4}$ inches tall. How much did he grow?

    Answer_____

14. Helen's daughter is $52\frac{5}{8}$ inches tall this year. Last year she was $49\frac{1}{8}$ inches tall. How much has she grown since last year?

    Answer_____

Changing Whole Numbers to Mixed Numbers

To change a whole number to a mixed number, you will need to borrow 1 from the whole number. Change the 1 to an improper fraction.

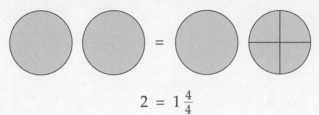

$$2 = 1\frac{4}{4}$$

Use These Steps

Change 4 to a mixed number with 6 as the denominator.

1. Borrow 1 from 4.

   $4 - 1 = 3$

2. Change 1 to an improper fraction with 6 as the denominator.

   $1 = \frac{6}{6}$

3. Write $\frac{6}{6}$ next to the 3.

   $3\frac{6}{6}$

Change each whole number to a mixed number with the given denominator.

1. $1 = \frac{\boxed{3}}{3}$
2. $1 = \frac{\Box}{5}$
3. $1 = \frac{\Box}{10}$
4. $1 = \frac{\Box}{7}$

5. $2 = 1\frac{\boxed{2}}{2}$
6. $3 = 2\frac{\Box}{5}$
7. $7 = 6\frac{\Box}{10}$
8. $9 = 8\frac{\Box}{12}$

9. $6 = 5\frac{\Box}{7}$
10. $4 = 3\frac{\Box}{6}$
11. $5 = 4\frac{\Box}{4}$
12. $8 = 7\frac{\Box}{9}$

13. $12 = 11\frac{\Box}{3}$
14. $15 = 14\frac{\Box}{16}$
15. $10 = 9\frac{\Box}{14}$
16. $17 = 16\frac{\Box}{15}$

17. $27 = \boxed{26}\frac{\boxed{5}}{5}$
18. $99 = \Box\frac{\Box}{12}$
19. $14 = \Box\frac{\Box}{17}$
20. $19 = \Box\frac{\Box}{20}$

21. $121 = \Box\frac{\Box}{8}$
22. $136 = \Box\frac{\Box}{11}$
23. $140 = \Box\frac{\Box}{18}$
24. $59 = \Box\frac{\Box}{21}$

25. $100 = \Box\frac{\Box}{25}$
26. $90 = \Box\frac{\Box}{30}$
27. $101 = \Box\frac{\Box}{49}$
28. $128 = \Box\frac{\Box}{100}$

Subtracting Fractions from One

To subtract a fraction from 1, you must change the 1 to an improper fraction. Use the denominator of the fraction being subtracted.

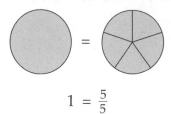

$$1 = \frac{5}{5}$$

Use These Steps

Subtract $1 - \frac{2}{5}$

1. Set up the problem.

$$\begin{array}{r} 1 \\ - \frac{2}{5} \\ \hline \end{array}$$

2. Change 1 to an improper fraction with 5 as the denominator. 5 is the denominator of the fraction being subtracted.

$$1 = \frac{5}{5}$$

3. Subtract the fractions.

$$\begin{array}{r} 1 = \frac{5}{5} \\ - \frac{2}{5} = \frac{2}{5} \\ \hline \frac{3}{5} \end{array}$$

Change 1 to an improper fraction with the given denominator.

1. $1 = \frac{\boxed{9}}{9}$
2. $1 = \frac{\boxed{}}{6}$
3. $1 = \frac{\boxed{}}{10}$
4. $1 = \frac{\boxed{}}{15}$
5. $1 = \frac{\boxed{}}{50}$

Subtract.

6. $1 - \frac{3}{7} =$

$$\begin{array}{r} 1 = \frac{7}{7} \\ - \frac{3}{7} = \frac{3}{7} \\ \hline \frac{4}{7} \end{array}$$

7. $1 - \frac{7}{12} =$

8. $1 - \frac{11}{15} =$

9. $1 - \frac{12}{25} =$

10. $\begin{array}{r} 1 \\ - \frac{5}{6} \\ \hline \end{array}$

11. $\begin{array}{r} 1 \\ - \frac{10}{19} \\ \hline \end{array}$

12. $\begin{array}{r} 1 \\ - \frac{6}{11} \\ \hline \end{array}$

13. $\begin{array}{r} 1 \\ - \frac{14}{17} \\ \hline \end{array}$

14. $\begin{array}{r} 1 \\ - \frac{29}{30} \\ \hline \end{array}$

15. $1 - \frac{2}{3} =$

16. $1 - \frac{9}{10} =$

17. $1 - \frac{3}{8} =$

18. $1 - \frac{7}{20} =$

Subtracting Fractions from Whole Numbers

To subtract a fraction from a whole number, you must borrow 1 from the whole number. Change 1 to an improper fraction. Use the denominator of the fraction being subtracted.

Use These Steps

Subtract $5 - \frac{1}{2}$

1. Set up the problem.

    $$\begin{array}{r} 5 \\ -\frac{1}{2} \\ \hline \end{array}$$

2. Borrow 1 from 5. $5 - 1 = 4$. Change the borrowed 1 to an improper fraction with 2 as the denominator.

    $$\begin{array}{r} 5 = 4\frac{2}{2} \\ -\frac{1}{2} = \frac{1}{2} \\ \hline \end{array}$$

3. Subtract the fractions. Bring down the whole number, 4.

    $$\begin{array}{r} 5 = 4\frac{2}{2} \\ -\frac{1}{2} = \frac{1}{2} \\ \hline 4\frac{1}{2} \end{array}$$

Subtract.

1. $2 - \frac{2}{3} =$

   $$\begin{array}{r} 2 = 1\frac{3}{3} \\ -\frac{2}{3} = \frac{2}{3} \\ \hline 1\frac{1}{3} \end{array}$$

2. $5 - \frac{3}{4} =$

3. $4 - \frac{3}{10} =$

4. $8 - \frac{4}{9} =$

5. $30 - \frac{11}{18}$

6. $26 - \frac{19}{20}$

7. $37 - \frac{13}{21}$

8. $40 - \frac{3}{25}$

9. $53 - \frac{12}{35}$

10. $10 - \frac{3}{5} =$

11. $13 - \frac{5}{6} =$

12. $15 - \frac{7}{12} =$

13. $21 - \frac{2}{15} =$

14. $94 - \frac{10}{49}$

15. $30 - \frac{18}{29}$

16. $76 - \frac{3}{50}$

17. $81 - \frac{37}{75}$

18. $25 - \frac{81}{100}$

Subtracting Fractions from Mixed Numbers

When you subtract a fraction from a mixed number, the fraction part of the mixed number may be smaller than the fraction being subtracted. When the whole number part is 1, change the 1 to an improper fraction.

Use These Steps

Subtract $1\frac{4}{9} - \frac{7}{9}$

1. Set up the problem.

   $1\frac{4}{9}$
   $-\ \frac{7}{9}$

2. Change 1 to an improper fraction with 9 as the denominator. Add $\frac{9}{9}$ to $\frac{4}{9}$.

   $1\frac{4}{9} = \frac{9}{9} + \frac{4}{9} = \frac{13}{9}$
   $-\ \frac{7}{9} = \qquad\qquad \frac{7}{9}$

3. Subtract from the new fraction, $\frac{13}{9}$. Reduce.

   $1\frac{4}{9} = \frac{13}{9}$
   $-\ \frac{7}{9} = \frac{7}{9}$
   $\qquad\quad \frac{6}{9} = \frac{2}{3}$

Subtract. Reduce if possible.

1. $1\frac{5}{12} - \frac{9}{12} =$

   $1\frac{5}{12} = \frac{12}{12} + \frac{5}{12} = \frac{17}{12}$
   $-\ \frac{9}{12} = \qquad\qquad\quad \frac{9}{12}$
   $\qquad\qquad\qquad\qquad\ \ \frac{8}{12} = \frac{2}{3}$

2. $1\frac{2}{7} - \frac{5}{7} =$

3. $1\frac{1}{4} - \frac{3}{4} =$

4. $1\frac{6}{11} - \frac{9}{11} =$

5. $1\frac{8}{15}$
   $-\ \frac{9}{15}$

6. $1\frac{10}{21}$
   $-\ \frac{20}{21}$

7. $1\frac{15}{32}$
   $-\ \frac{19}{32}$

8. $1\frac{13}{50}$
   $-\ \frac{29}{50}$

9. $1\frac{15}{63}$
   $-\ \frac{30}{63}$

10. $1\frac{11}{18} - \frac{17}{18} =$

11. $1\frac{29}{42} - \frac{31}{42} =$

12. $1\frac{15}{76} - \frac{40}{76} =$

13. $1\frac{26}{100} - \frac{37}{100} =$

14. $1\frac{2}{7}$
    $-\ \frac{3}{7}$

15. $1\frac{7}{10}$
    $-\ \frac{9}{10}$

16. $1\frac{5}{16}$
    $-\ \frac{9}{16}$

17. $1\frac{11}{25}$
    $-\ \frac{21}{25}$

18. $1\frac{23}{54}$
    $-\ \frac{40}{54}$

Subtracting Fractions from Mixed Numbers

When you subtract a fraction from a mixed number, you may have to borrow from the whole number. Change the borrowed 1 to an improper fraction. Use the denominator of the fraction being subtracted.

Use These Steps

Subtract $4\frac{1}{10} - \frac{3}{10}$

1. Since $\frac{3}{10}$ is greater than $\frac{1}{10}$, borrow 1 from the 4. Change the borrowed 1 to an improper fraction with 10 as the denominator. $1 = \frac{10}{10}$.

   $4\frac{1}{10} = 3\frac{1}{10} + \frac{10}{10}$

2. Add the mixed number and the fraction.

   $3\frac{1}{10} + \frac{10}{10} = 3\frac{11}{10}$

3. Subtract the fractions. Bring down the whole number. Reduce.

   $3\frac{11}{10} - \frac{3}{10} = 3\frac{8}{10} = 3\frac{4}{5}$

Subtract. Reduce if possible.

1. $11\frac{5}{11} = 10\frac{16}{11}$
   $-\frac{6}{11} = \frac{6}{11}$
   $= 10\frac{10}{11}$

2. $31\frac{2}{8} - \frac{5}{8}$

3. $12\frac{1}{6} - \frac{5}{6}$

4. $20\frac{4}{12} - \frac{7}{12}$

5. $45\frac{3}{15} - \frac{11}{15}$

6. $7\frac{2}{9} - \frac{4}{9} =$
   $7\frac{2}{9} = 6\frac{11}{9}$
   $-\frac{4}{9} = \frac{4}{9}$
   $= 6\frac{7}{9}$

7. $4\frac{3}{10} - \frac{7}{10} =$

8. $6\frac{1}{4} - \frac{3}{4} =$

9. $2\frac{3}{7} - \frac{6}{7} =$

10. $17\frac{2}{21} - \frac{3}{21} =$

11. $40\frac{9}{18} - \frac{13}{18} =$

12. $13\frac{7}{30} - \frac{11}{30} =$

13. $72\frac{17}{50} - \frac{20}{50} =$

14. $18\frac{2}{63} - \frac{4}{63}$

15. $25\frac{18}{72} - \frac{27}{72}$

16. $31\frac{25}{75} - \frac{70}{75}$

17. $6\frac{3}{90} - \frac{43}{90}$

18. $12\frac{12}{100} - \frac{92}{100}$

Subtracting Mixed Numbers from Mixed Numbers

When subtracting a mixed number from a mixed number, the fraction on top may be smaller than the fraction on the bottom. When this happens, you will need to borrow from the whole number on top.

Use These Steps

Subtract $6\frac{2}{7} - 3\frac{4}{7}$

1. Set up the problem. Since $\frac{4}{7}$ is greater than $\frac{2}{7}$, borrow 1 from the 6. Change the borrowed 1 to an improper fraction with 7 as the denominator. $1 = \frac{7}{7}$.

    $6\frac{2}{7} = 5\frac{2}{7} + \frac{7}{7}$
    $- 3\frac{4}{7}$

2. Add $5\frac{2}{7}$ to $\frac{7}{7}$.

    $5\frac{2}{7} + \frac{7}{7} = 5\frac{9}{7}$

3. Subtract the fractions. Then subtract the whole numbers.

    $5\frac{9}{7}$
    $- 3\frac{4}{7}$
    $\overline{2\frac{5}{7}}$

Subtract. Reduce if possible.

1. $8\frac{1}{3} - 4\frac{2}{3} =$

    $8\frac{1}{3} = 7\frac{4}{3}$
    $-4\frac{2}{3} = 4\frac{2}{3}$
    $\overline{3\frac{2}{3}}$

2. $5\frac{1}{4} - 3\frac{3}{4} =$

3. $6\frac{5}{7} - 1\frac{6}{7} =$

4. $3\frac{1}{5} - 1\frac{2}{5} =$

5. $10\frac{3}{8}$
   $- 5\frac{7}{8}$

6. $13\frac{1}{6}$
   $- 12\frac{3}{6}$

7. $19\frac{2}{10}$
   $- 9\frac{3}{10}$

8. $25\frac{10}{15}$
   $- 15\frac{13}{15}$

9. $30\frac{6}{19}$
   $- 24\frac{10}{19}$

10. Lloyd began his trip with $18\frac{1}{4}$ gallons of gas in the tank of his car. He used $16\frac{3}{4}$ gallons. How many gallons does he have left in his gas tank?

    Answer _____

11. Marsha's high-jump record last year was $52\frac{7}{10}$ inches. This year she jumped $59\frac{3}{10}$ inches. How much higher was her record this year?

    Answer _____

Real-Life Application — Time Off

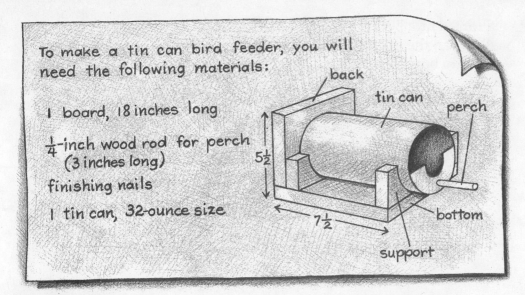

Example The supports for the tin can are each $2\frac{1}{2}$ inches high. How much wood is left after cutting off enough for one support?

$$18 = 17\frac{2}{2}$$
$$-2\frac{1}{2} = 2\frac{1}{2}$$
$$\overline{\phantom{-2\frac{1}{2} = }\;15\frac{1}{2}}$$

There are $15\frac{1}{2}$ inches of wood left.

Solve.

1. You need a second support for the feeder. You have already cut the first support from the 18-inch long board. How much board will you have left after you cut off the second support?

   Answer_____

2. The bottom of the feeder is $7\frac{1}{2}$ inches long. If you have a board that is 13 inches long, how much wood is left after cutting off enough for the bottom?

   Answer_____

3. The perch should be $2\frac{3}{4}$ inches long. How far into the tin can should you place the 3-inch long rod?

   Answer_____

4. The tin can shown is 8 inches long. How much longer than the bottom of the feeder is the tin can?

   Answer_____

Mixed Review

Add or subtract. Reduce if possible.

1. $\dfrac{1}{6} + \dfrac{5}{6} =$
2. $7\dfrac{2}{5} - \dfrac{2}{5} =$
3. $2\dfrac{5}{8} - 1\dfrac{1}{8} =$
4. $\dfrac{2}{3} + \dfrac{2}{3} =$

5. $6\dfrac{3}{4} - 2\dfrac{1}{4}$
6. $9\dfrac{1}{2} + 3\dfrac{1}{2}$
7. $12\dfrac{5}{7} - \dfrac{2}{7}$
8. $\dfrac{5}{12} + \dfrac{7}{12}$
9. $\dfrac{15}{16} - \dfrac{3}{16}$

Change each whole number to a mixed number with the given denominator.

10. $9 = 8\dfrac{\square}{2}$
11. $6 = 5\dfrac{\square}{3}$
12. $18 = \square\dfrac{\square}{9}$
13. $20 = \square\dfrac{\square}{10}$
14. $29 = \square\dfrac{\square}{35}$

Add or subtract. Reduce if possible.

15. $1 - \dfrac{3}{4} =$
16. $2 - \dfrac{2}{5} =$
17. $14 + \dfrac{17}{20} =$
18. $20 - \dfrac{3}{16} =$

19. $12\dfrac{3}{10} - 7\dfrac{7}{10}$
20. $17\dfrac{1}{6} + 3\dfrac{5}{6}$
21. $7\dfrac{9}{11} - 2\dfrac{10}{11}$
22. $1\dfrac{13}{15} + \dfrac{14}{15}$
23. $8\dfrac{2}{9} - 3\dfrac{7}{9}$

24. $24\dfrac{5}{12} + 6\dfrac{1}{12} =$
25. $13\dfrac{5}{16} - 12\dfrac{9}{16} =$
26. $6\dfrac{3}{4} + 1\dfrac{1}{4} =$
27. $19\dfrac{2}{21} - \dfrac{4}{21} =$

28. $1\dfrac{1}{3} - \dfrac{2}{3}$
29. $2\dfrac{31}{50} - 1\dfrac{41}{50}$
30. $20\dfrac{7}{32} + 11\dfrac{5}{32}$
31. $1\dfrac{15}{64} - \dfrac{7}{64}$
32. $27\dfrac{23}{100} - \dfrac{17}{100}$

Subtracting Fractions with Different Denominators

When subtracting fractions, the denominators may be different. When this happens, you need to raise fractions to higher terms with the same common denominator. Then subtract the numerators.

When one denominator will divide evenly into the other, use the larger denominator as the common denominator.

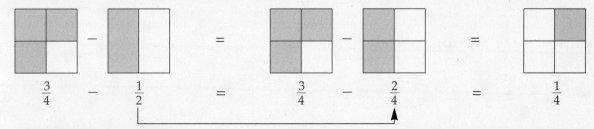

$$\frac{3}{4} - \frac{1}{2} = \frac{3}{4} - \frac{2}{4} = \frac{1}{4}$$

Use These Steps

Subtract $\frac{5}{8} - \frac{1}{4}$

1. The denominators are different. Since 4 divides evenly into 8, write 8 as the common denominator.

   $$\frac{1}{4} = \frac{}{4 \times 2} = \frac{}{8}$$

2. Raise $\frac{1}{4}$ to higher terms with 8 as the denominator.

   $$\frac{1}{4} = \frac{1 \times 2}{4 \times 2} = \frac{2}{8}$$

3. Subtract the numerators. Write the difference over the denominator.

   $$\frac{5}{8} = \frac{5}{8}$$
   $$-\frac{1}{4} = \frac{2}{8}$$
   $$\frac{3}{8}$$

Subtract. Reduce if possible.

1. $\frac{9}{10} - \frac{1}{5} =$
$$\frac{9}{10} = \frac{9}{10}$$
$$-\frac{1}{5} = \frac{2}{10}$$
$$\frac{7}{10}$$

2. $\frac{8}{12} - \frac{1}{4} =$

3. $\frac{2}{3} - \frac{2}{9} =$

4. $\frac{5}{7} - \frac{1}{14} =$

5. $\frac{7}{15}$
$-\frac{1}{5}$

6. $\frac{5}{9}$
$-\frac{5}{18}$

7. $\frac{9}{10}$
$-\frac{7}{20}$

8. $\frac{3}{4}$
$-\frac{9}{16}$

9. $\frac{11}{20}$
$-\frac{2}{5}$

10. $\frac{17}{18} - \frac{5}{6} =$

11. $\frac{7}{10} - \frac{1}{2} =$

12. $\frac{9}{20} - \frac{1}{4} =$

13. $\frac{11}{14} - \frac{1}{2} =$

Subtracting Fractions with Different Denominators

Find a common denominator. Subtract only the numerators.
Reduce the fraction if possible.

Use These Steps

Subtract $\frac{9}{10} - \frac{2}{5}$

1. Raise $\frac{2}{5}$ to higher terms with 10 as the denominator.

$$\frac{2}{5} = \frac{2 \times 2}{5 \times 2} = \frac{4}{10}$$

2. Subtract the numerators. Write the difference over the denominator.

$$\frac{9}{10} = \frac{9}{10}$$
$$-\frac{2}{5} = \frac{4}{10}$$
$$\frac{5}{10}$$

3. Reduce the answer to lowest terms.

$$\frac{5}{10} = \frac{1}{2}$$

Subtract. Reduce if possible.

1.
$$\frac{7}{12} - \frac{1}{3} =$$
$$\frac{7}{12} = \frac{7}{12}$$
$$-\frac{1}{3} = \frac{4}{12}$$
$$\frac{3}{12} = \frac{1}{4}$$

2. $\frac{1}{2} - \frac{1}{8} =$

3. $\frac{2}{3} - \frac{1}{9} =$

4. $\frac{11}{14} - \frac{2}{7} =$

5. $\frac{3}{4} - \frac{3}{16}$

6. $\frac{19}{20} - \frac{1}{5}$

7. $\frac{11}{18} - \frac{1}{3}$

8. $\frac{5}{7} - \frac{8}{21}$

9. $\frac{19}{24} - \frac{1}{6}$

10. $\frac{3}{5} - \frac{2}{25} =$

11. $\frac{24}{27} - \frac{2}{9} =$

12. $\frac{29}{30} - \frac{5}{6} =$

13. $\frac{5}{8} - \frac{7}{40} =$

14. Janie gave her son $\frac{1}{2}$ teaspoon of cough medicine. She gave her daughter $\frac{1}{4}$ teaspoon. How much more medicine did she give her son?

15. Viola swam $\frac{4}{5}$ mile. Rhonda swam $\frac{7}{10}$ mile. How much farther did Viola swim?

Answer _____

Answer _____

Subtracting Fractions with Different Denominators

When you subtract fractions, you can find a common denominator by multiplying the denominators. Make sure that each of the old denominators divides into the common denominator. Then raise each fraction to higher terms with the new denominator.

Use These Steps

Subtract $\frac{1}{3} - \frac{1}{4}$

1. Multiply 3 times 4 to find a common denominator. $3 \times 4 = 12$.

$$\frac{1}{3} = \frac{}{12}$$
$$-\frac{1}{4} = \frac{}{12}$$

2. Raise each fraction to higher terms with 12 as the denominator.

$$\frac{1}{3} = \frac{4}{12}$$
$$-\frac{1}{4} = \frac{3}{12}$$

3. Subtract the numerators. Write the difference over the denominator.

$$\frac{1}{3} = \frac{4}{12}$$
$$-\frac{1}{4} = \frac{3}{12}$$
$$\frac{1}{12}$$

Subtract.

1. $\frac{3}{5} - \frac{1}{2} =$

$\frac{3}{5} = \frac{6}{10}$
$-\frac{1}{2} = \frac{5}{10}$
$\phantom{-\frac{1}{2} =}\frac{1}{10}$

2. $\frac{3}{4} - \frac{1}{3} =$

3. $\frac{2}{3} - \frac{1}{2} =$

4. $\frac{4}{5} - \frac{2}{3} =$

5. $\frac{5}{6}$
$-\frac{1}{5}$

6. $\frac{3}{8}$
$-\frac{1}{3}$

7. $\frac{4}{7}$
$-\frac{1}{2}$

8. $\frac{2}{3}$
$-\frac{3}{10}$

9. $\frac{3}{4}$
$-\frac{2}{5}$

10. $\frac{3}{7} - \frac{1}{6} =$

11. $\frac{2}{3} - \frac{2}{7} =$

12. $\frac{5}{7} - \frac{1}{2} =$

13. $\frac{8}{9} - \frac{3}{4} =$

14. $\frac{4}{7}$
$-\frac{2}{5}$

15. $\frac{7}{9}$
$-\frac{1}{2}$

16. $\frac{3}{4}$
$-\frac{1}{5}$

17. $\frac{3}{5}$
$-\frac{1}{4}$

18. $\frac{5}{8}$
$-\frac{1}{3}$

Subtracting Fractions Using the LCD

To find the lowest common denominator (LCD), list several multiples of each denominator. Find the smallest number that appears on both lists. This number is the LCD. Make sure that each of the old denominators divides evenly into the new denominator. Then subtract.

Use These Steps

Subtract $\frac{3}{8} - \frac{1}{6}$

1. List multiples of each denominator. The smallest number on both lists is 24.

   $\frac{3}{8}$ 8 16 ⟨24⟩ 32 40

   $\frac{1}{6}$ 6 12 18 ⟨24⟩ 30

2. Raise each fraction to higher terms with 24 as the LCD.

   $$\frac{3}{8} = \frac{9}{24}$$
   $$-\frac{1}{6} = \frac{4}{24}$$

3. Subtract the numerators. Write the difference over the denominator.

   $$\frac{3}{8} = \frac{9}{24}$$
   $$-\frac{1}{6} = \frac{4}{24}$$
   $$\frac{5}{24}$$

Subtract. Use the LCD.

1. $\frac{3}{4} - \frac{1}{6} =$

   $\frac{3}{4}$ 4 8 ⟨12⟩ 16

   $\frac{1}{6}$ 6 ⟨12⟩ 18

   $$\frac{3}{4} = \frac{9}{12}$$
   $$-\frac{1}{6} = \frac{2}{12}$$
   $$\frac{7}{12}$$

2. $\frac{5}{6} - \frac{4}{9} =$

3. $\frac{7}{10} - \frac{1}{4} =$

4. $\frac{7}{8} - \frac{7}{12} =$

5. $\frac{13}{18} - \frac{5}{12}$

6. $\frac{9}{10} - \frac{2}{3}$

7. $\frac{7}{9} - \frac{7}{12}$

8. $\frac{14}{15} - \frac{5}{9}$

9. $\frac{11}{12} - \frac{3}{16}$

Mixed Review

Subtract. Reduce if possible.

1. $\frac{4}{5} - \frac{3}{8} =$
2. $\frac{5}{6} - \frac{1}{8} =$
3. $\frac{5}{8} - \frac{1}{2} =$
4. $\frac{3}{4} - \frac{3}{7} =$

5. $\frac{9}{10} - \frac{5}{8} =$
6. $\frac{4}{7} - \frac{2}{5} =$
7. $\frac{3}{10} - \frac{1}{5} =$
8. $\frac{2}{3} - \frac{5}{14} =$

9. $\frac{6}{7} - \frac{1}{2}$
10. $\frac{10}{13} - \frac{1}{3}$
11. $\frac{7}{8} - \frac{7}{12}$
12. $\frac{7}{12} - \frac{1}{6}$
13. $\frac{2}{3} - \frac{1}{9}$

14. $\frac{2}{3} - \frac{2}{9}$
15. $\frac{7}{10} - \frac{1}{4}$
16. $\frac{3}{4} - \frac{1}{6}$
17. $\frac{5}{9} - \frac{1}{2}$
18. $\frac{5}{6} - \frac{3}{5}$

19. $\frac{14}{15} - \frac{5}{9} =$
20. $\frac{13}{18} - \frac{5}{12} =$
21. $\frac{1}{4} - \frac{1}{5} =$
22. $\frac{19}{20} - \frac{3}{4} =$

23. $\frac{5}{6} - \frac{3}{4} =$
24. $\frac{2}{3} - \frac{1}{10} =$
25. $\frac{11}{12} - \frac{4}{9} =$
26. $\frac{13}{15} - \frac{1}{5} =$

Problem Solving: Using a Map

The map shows the distance in miles between four towns. For example, the distance between Prairie and Keene is $6\frac{3}{10}$ miles.

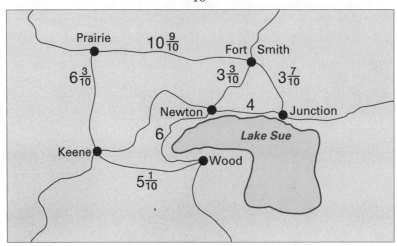

Example Bob sells insurance. Bob is traveling from Prairie to Fort Smith. After driving for $6\frac{4}{10}$ miles he stopped for a cup of coffee. How many more miles does he have to drive to Fort Smith?

▶ **Step 1.** Find the distance from Prairie to Fort Smith on the map.

$$10\frac{9}{10} \text{ miles}$$

▶ **Step 2.** To find how many more miles Bob has to drive, subtract. Reduce.

$$\begin{array}{r} 10\frac{9}{10} \\ -\ 6\frac{4}{10} \\ \hline 4\frac{5}{10} = 4\frac{1}{2} \end{array}$$

Bob has to drive $4\frac{1}{2}$ more miles to Fort Smith.

Solve. Reduce if possible.

1. The distance from Fort Smith to Newton is $3\frac{3}{10}$ miles. How much farther is Junction from Fort Smith than Newton is?

2. The distance from Fort Smith to Keene is 12 miles. The distance from Fort Smith to Newton is $3\frac{3}{10}$ miles. How far is it from Newton to Keene?

Answer_____ Answer_____

Solve. Reduce if possible.

3. The distance from Fort Smith to Keene is how many more miles than the distance from Fort Smith to Prairie?

   Answer_____

4. One day the road from Fort Smith to Newton was closed. Bob took a detour through Junction. How far did he travel to get to Newton?

   Answer_____

5. Bob knows a short cut from Newton to Prairie. The short cut is $7\frac{1}{10}$ miles long. The long way is 15 miles long. How many miles will he save if he uses the short cut instead of traveling through Keene?

   Answer_____

6. The distance across Lake Sue from Junction to Wood is $5\frac{1}{2}$ miles. How much shorter is it to cross the lake by boat than to go the 10 miles around the lake to Wood?

   Answer_____

7. How much farther is it from Wood to Newton than it is from Wood to Keene?

   Answer_____

8. Bob's sister's house is $2\frac{1}{10}$ miles past Keene on the way to Wood. Does she live closer to Wood or to Keene?

   Answer_____

9. Bob's car broke down one day $2\frac{3}{10}$ miles past Keene on the way to Wood. How many miles did a tow truck from Wood have to come to pick up Bob's car?

   Answer_____

10. Bob drove $42\frac{1}{10}$ miles last week. This week he drove $3\frac{7}{10}$ miles less than last week. How many miles did he drive this week?

    Answer_____

99

Subtracting Mixed Numbers

When subtracting mixed numbers, you may need to find a common denominator for the fractions before you can subtract.

Use These Steps

Subtract $3\frac{1}{2} - 1\frac{1}{3}$

1. Find a common denominator by multiplying the two denominators.
$2 \times 3 = 6$.

$$3\frac{1}{2} = 3\frac{}{6}$$
$$-1\frac{1}{3} = 1\frac{}{6}$$

2. Raise each fraction to higher terms with 6 as the denominator.

$$3\frac{1}{2} = 3\frac{3}{6}$$
$$-1\frac{1}{3} = 1\frac{2}{6}$$

3. Subtract the fractions. Subtract the whole numbers.

$$3\frac{1}{2} = 3\frac{3}{6}$$
$$-1\frac{1}{3} = 1\frac{2}{6}$$
$$\phantom{-1\frac{1}{3} = }2\frac{1}{6}$$

Subtract. Reduce if possible.

1. $4\frac{3}{8} - 2\frac{1}{4} =$

    $4\frac{3}{8} = 4\frac{3}{8}$
    $-2\frac{1}{4} = 2\frac{2}{8}$
    $\phantom{-2\frac{1}{4} = }2\frac{1}{8}$

2. $7\frac{2}{3} - 3\frac{1}{6} =$

3. $5\frac{3}{4} - 2\frac{1}{2} =$

4. $9\frac{3}{5} - 7\frac{1}{10} =$

5. $12\frac{3}{4}$
    $-2\frac{1}{3}$

6. $15\frac{5}{6}$
    $-10\frac{3}{5}$

7. $20\frac{7}{10}$
    $-8\frac{1}{3}$

8. $13\frac{5}{9}$
    $-1\frac{1}{2}$

9. $12\frac{3}{8}$
    $-5\frac{1}{4}$

10. Kho bought $3\frac{3}{4}$ yards of blue material to make a small quilt. She used $1\frac{2}{3}$ yards for blocks and the rest for the border. How much material did she use for the border?

11. Kho bought $4\frac{7}{8}$ yards of green material for the quilt. She used $1\frac{2}{3}$ yards for blocks and the rest for the lining. How much did she use for the lining?

Answer _____   Answer _____

Subtracting Mixed Numbers with Borrowing

When subtracting mixed numbers, you may need to find a common denominator. Then, if the fraction you are subtracting is greater than the one you are subtracting from, you will need to borrow.

Use These Steps

Subtract $18\frac{2}{7} - 10\frac{1}{2}$

1. Find a common denominator. Raise each fraction to higher terms.

$$18\frac{2}{7} = 18\frac{4}{14}$$
$$-10\frac{1}{2} = 10\frac{7}{14}$$

2. Since $\frac{7}{14}$ is greater than $\frac{4}{14}$, borrow 1 from 18.

$$18\frac{2}{7} = 18\frac{4}{14} = 17\frac{18}{14}$$
$$-10\frac{1}{2} = 10\frac{7}{14} = 10\frac{7}{14}$$

3. Subtract the fractions. Subtract the whole numbers.

$$17\frac{18}{14}$$
$$-10\frac{7}{14}$$
$$7\frac{11}{14}$$

Subtract. Reduce if possible.

1. $7\frac{2}{3} - 6\frac{8}{9} =$

$$7\frac{2}{3} = 7\frac{6}{9} = 6\frac{15}{9}$$
$$-6\frac{8}{9} = 6\frac{8}{9} = 6\frac{8}{9}$$
$$\phantom{-6\frac{8}{9} = 6\frac{8}{9} = 6}\frac{7}{9}$$

2. $9\frac{1}{2} - 6\frac{7}{8} =$

3. $10\frac{1}{6} - 2\frac{3}{8} =$

4. $27\frac{2}{9}$
$-18\frac{3}{4}$

5. $30\frac{1}{10}$
$-15\frac{3}{5}$

6. $14\frac{3}{8}$
$-6\frac{11}{12}$

7. $33\frac{2}{11} - 19\frac{1}{2} =$

8. $25\frac{1}{5} - 9\frac{1}{3} =$

9. $18\frac{3}{10} - 17\frac{3}{4} =$

10. $50\frac{9}{25}$
$-14\frac{4}{5}$

11. $86\frac{5}{9}$
$-49\frac{5}{6}$

12. $132\frac{9}{20}$
$-75\frac{3}{4}$

Subtracting Mixed Numbers with Borrowing

When subtracting mixed numbers with different denominators, change the fractions to higher terms with common denominators. Then borrow.

Use These Steps

Subtract $2\frac{1}{5} - 1\frac{4}{15}$

1. Find a common denominator. Raise each fraction to higher terms.

   $2\frac{1}{5} = 2\frac{3}{15}$
   $-1\frac{4}{15} = 1\frac{4}{15}$

2. Since $\frac{4}{15}$ is greater than $\frac{3}{15}$, borrow 1 from the 2.

   $2\frac{1}{5} = 2\frac{3}{15} = 1\frac{18}{15}$
   $-1\frac{4}{15} = 1\frac{4}{15} = 1\frac{4}{15}$

3. Subtract the fractions. Subtract the whole numbers.

   $1\frac{18}{15}$
   $-1\frac{4}{15}$
   $\frac{14}{15}$

Subtract. Reduce if possible.

1. $6\frac{1}{10} - 2\frac{1}{6} =$

   $6\frac{1}{10} = 6\frac{3}{30} = 5\frac{33}{30}$
   $-2\frac{1}{6} = 2\frac{5}{30} = 2\frac{5}{30}$
   $\phantom{-2\frac{1}{6} = 2\frac{5}{30} =\ } 3\frac{28}{30} = 3\frac{14}{15}$

2. $4\frac{2}{7} - 1\frac{1}{3} =$

3. $9\frac{2}{5} - 3\frac{1}{2} =$

4. $12\frac{1}{12} - 5\frac{5}{6} =$

5. $10\frac{3}{8} - 9\frac{5}{6} =$

6. $15\frac{1}{10} - 11\frac{2}{5} =$

7. $20\frac{3}{20} - 7\frac{1}{4} =$

8. $9\frac{1}{7} - 3\frac{4}{21} =$

9. $14\frac{1}{6} - 8\frac{7}{30} =$

Subtracting Mixed Numbers and Fractions

When the fraction you are subtracting is greater than the fraction you are subtracting from, you will need to borrow.

Use These Steps

Subtract $1\frac{1}{3} - \frac{1}{2}$

1. Find a common denominator.

$$1\frac{1}{3} = 1\frac{2}{6}$$
$$-\frac{1}{2} = \frac{3}{6}$$

2. Since $\frac{3}{6}$ is greater than $\frac{2}{6}$, borrow 1.

$$1\frac{1}{3} = 1\frac{2}{6} = \frac{8}{6}$$
$$-\frac{1}{2} = \frac{3}{6} = \frac{3}{6}$$

3. Subtract the fractions.

$$\frac{8}{6}$$
$$-\frac{3}{6}$$
$$\frac{5}{6}$$

Subtract. Reduce if possible.

1.
$$1\frac{1}{10} - \frac{5}{6} =$$
$$1\frac{1}{10} = 1\frac{3}{30} = \frac{33}{30}$$
$$-\frac{5}{6} = \frac{25}{30} = \frac{25}{30}$$
$$\frac{8}{30} = \frac{4}{15}$$

2. $3\frac{2}{5} - \frac{7}{8} =$

3. $1\frac{3}{7} - \frac{3}{4} =$

4. $4\frac{1}{2} - \frac{2}{3} =$

5. $9\frac{3}{8} - \frac{1}{2} =$

6. $1\frac{1}{9} - \frac{1}{6} =$

7. $5\frac{2}{5} - \frac{9}{10} =$

8. $10\frac{5}{12} - \frac{5}{6} =$

9. $2\frac{3}{11} - \frac{1}{3} =$

Unit 3 Review

Subtract. Reduce if possible.

1. $\dfrac{11}{13} - \dfrac{2}{13}$

2. $\dfrac{7}{10} - \dfrac{1}{10}$

3. $\dfrac{5}{8} - \dfrac{3}{8}$

4. $\dfrac{4}{5} - \dfrac{3}{5}$

5. $\dfrac{5}{9} - \dfrac{2}{9}$

6. $\dfrac{10}{15} - \dfrac{5}{15} =$

7. $\dfrac{5}{6} - \dfrac{1}{6} =$

8. $\dfrac{19}{21} - \dfrac{12}{21} =$

9. $\dfrac{17}{30} - \dfrac{2}{30} =$

10. $12\dfrac{3}{10} - 4\dfrac{1}{10}$

11. $7\dfrac{3}{4} - 5\dfrac{1}{4}$

12. $25\dfrac{9}{16} - 11\dfrac{3}{16}$

13. $19\dfrac{14}{18} - 16\dfrac{2}{18}$

14. $8\dfrac{7}{20} - 2\dfrac{3}{20}$

15. $10\dfrac{7}{8} - \dfrac{5}{8} =$

16. $4\dfrac{15}{25} - 3 =$

17. $18\dfrac{30}{35} - \dfrac{18}{35} =$

18. $2\dfrac{22}{40} - \dfrac{7}{40} =$

19. $5\dfrac{4}{30} - \dfrac{1}{30}$

20. $8\dfrac{20}{27} - \dfrac{2}{27}$

21. $60\dfrac{47}{50} - 30$

22. $5\dfrac{17}{42} - \dfrac{10}{42}$

23. $14\dfrac{10}{11} - 6$

24. $1 - \dfrac{5}{6} =$

25. $1 - \dfrac{7}{10} =$

26. $1 - \dfrac{3}{4} =$

27. $1 - \dfrac{1}{3} =$

Subtract. Reduce if possible.

28. $4 - \frac{1}{2} =$

29. $6 - \frac{1}{3} =$

30. $3 - \frac{3}{4} =$

31. $6 - \frac{2}{9} =$

32. $\begin{array}{r} 8 \\ - \frac{3}{10} \\ \hline \end{array}$

33. $\begin{array}{r} 10 \\ - \frac{2}{5} \\ \hline \end{array}$

34. $\begin{array}{r} 18 \\ - \frac{1}{6} \\ \hline \end{array}$

35. $\begin{array}{r} 11 \\ - \frac{4}{7} \\ \hline \end{array}$

36. $\begin{array}{r} 13 \\ - \frac{5}{12} \\ \hline \end{array}$

37. $5\frac{1}{3} - 1\frac{2}{3} =$

38. $1\frac{4}{9} - \frac{7}{9} =$

39. $7\frac{1}{4} - \frac{3}{4} =$

40. $12\frac{4}{11} - 3\frac{5}{11} =$

41. $\begin{array}{r} 1\frac{5}{9} \\ - \frac{7}{9} \\ \hline \end{array}$

42. $\begin{array}{r} 8\frac{3}{10} \\ - 2\frac{7}{10} \\ \hline \end{array}$

43. $\begin{array}{r} 2\frac{1}{5} \\ - \frac{3}{5} \\ \hline \end{array}$

44. $\begin{array}{r} 1\frac{1}{6} \\ - \frac{3}{6} \\ \hline \end{array}$

45. $\begin{array}{r} 10\frac{3}{20} \\ - 5\frac{7}{20} \\ \hline \end{array}$

46. $\frac{2}{3} - \frac{1}{6} =$

47. $\frac{3}{7} - \frac{5}{14} =$

48. $\frac{11}{12} - \frac{3}{4} =$

49. $\frac{4}{5} - \frac{8}{20} =$

50. $\begin{array}{r} \frac{1}{2} \\ - \frac{1}{3} \\ \hline \end{array}$

51. $\begin{array}{r} \frac{3}{4} \\ - \frac{3}{5} \\ \hline \end{array}$

52. $\begin{array}{r} \frac{9}{10} \\ - \frac{2}{3} \\ \hline \end{array}$

53. $\begin{array}{r} \frac{3}{7} \\ - \frac{1}{6} \\ \hline \end{array}$

54. $\begin{array}{r} \frac{7}{8} \\ - \frac{2}{3} \\ \hline \end{array}$

55. $\frac{3}{4} - \frac{1}{6} =$

56. $\frac{3}{10} - \frac{1}{4} =$

57. $\frac{5}{8} - \frac{1}{12} =$

58. $\frac{7}{10} - \frac{3}{8} =$

Subtract. Reduce if possible.

59. $5\frac{1}{3} - 2\frac{1}{4}$

60. $10\frac{3}{5} - 7\frac{1}{10}$

61. $13\frac{5}{6} - 2\frac{1}{4}$

62. $8\frac{2}{3} - 7\frac{5}{12}$

63. $6\frac{8}{15} - 1\frac{1}{2}$

64. $21\frac{3}{10} - 2\frac{1}{4} =$

65. $18\frac{7}{9} - 13\frac{1}{3} =$

66. $10\frac{3}{4} - 5\frac{1}{6} =$

67. $9\frac{7}{15} - 2\frac{2}{5} =$

68. $15\frac{3}{7} - 6\frac{1}{2}$

69. $4\frac{1}{2} - 2\frac{5}{8}$

70. $16\frac{1}{10} - 3\frac{3}{4}$

71. $10\frac{1}{9} - 6\frac{3}{4} =$

72. $31\frac{5}{11} - 18\frac{2}{3} =$

73. $17\frac{1}{6} - 3\frac{3}{5} =$

74. $1\frac{2}{9} - \frac{1}{4}$

75. $5\frac{2}{5} - \frac{5}{8}$

76. $2\frac{1}{12} - \frac{3}{10}$

Below is a list of the problems in this review and the pages on which the skills are taught. If you missed any problems, turn to the pages listed and practice the skills. Then correct the problems you missed in the Unit Review.

Problems	Pages
1-14	75-78
15-23	80-81
24-27	86
28-36	87
37-45	88-90

Problems	Pages
46-58	93-96
59-67	100
68-73	101-102
74-76	103

Unit 4 MULTIPLYING FRACTIONS

Have you ever had to double a recipe, change a fraction of a yard to inches, or figure the cost of clothes at a $\frac{1}{3}$-off sale? All of these problems can be solved by multiplying fractions.

In this unit you will learn how to multiply fractions, whole numbers, and mixed numbers. You will be able to solve word problems using fractions.

Getting Ready

You should be familiar with the skills on this page and the next before you begin this unit. To check your answers, turn to page 193.

 Knowing the multiplication facts will help make multiplying fractions easier.

Multiply.

1. $3 \times 4 = $ **12**
2. $6 \times 2 = $
3. $1 \times 5 = $
4. $2 \times 9 = $

5. $7 \times 3 = $
6. $4 \times 5 = $
7. $3 \times 2 = $
8. $5 \times 8 = $

9. $9 \times 3 = $
10. $3 \times 3 = $
11. $6 \times 7 = $
12. $4 \times 8 = $

13. $7 \times 9 = $
14. $5 \times 3 = $
15. $2 \times 5 = $
16. $4 \times 4 = $

For review, see pages 85–87 in Math Matters for Adults, Whole Numbers.

Getting Ready

 Often after multiplying fractions, you need to reduce answers to lowest terms.

Reduce each fraction to lowest terms. If the fraction is in lowest terms, write LT.

17. $\frac{2}{12} = \frac{1}{6}$
18. $\frac{6}{20} =$
19. $\frac{4}{63} =$
20. $\frac{18}{32} =$
21. $\frac{3}{9} =$

22. $\frac{12}{28} =$
23. $\frac{15}{32} =$
24. $\frac{16}{20} =$
25. $\frac{12}{45} =$
26. $\frac{14}{42} =$

27. $\frac{20}{25} =$
28. $\frac{9}{10} =$
29. $\frac{2}{14} =$
30. $\frac{8}{16} =$
31. $\frac{10}{50} =$

For review, see Unit 1, pages 15–16.

 When you get an improper fraction as an answer, you need to change the improper fraction to a whole or mixed number.

Change each improper fraction to a whole or mixed number. Reduce if possible.

32. $\frac{11}{2} =$

$$2\overline{)11} = 5\frac{1}{2}$$
$$\underline{-10}$$
$$1$$

33. $\frac{16}{5} =$
34. $\frac{72}{4} =$
35. $\frac{29}{3} =$
36. $\frac{51}{6} =$

37. $\frac{40}{12} =$
38. $\frac{27}{8} =$
39. $\frac{42}{7} =$
40. $\frac{49}{9} =$
41. $\frac{53}{10} =$

42. $\frac{18}{3} =$
43. $\frac{22}{11} =$
44. $\frac{25}{4} =$
45. $\frac{30}{14} =$
46. $\frac{21}{20} =$

47. $\frac{50}{16} =$
48. $\frac{96}{24} =$
49. $\frac{92}{8} =$
50. $\frac{45}{15} =$
51. $\frac{54}{13} =$

For review, see Unit 1, pages 33–34.

Multiplying Fractions by Fractions

Multiplying fractions is a way to find a part of something. If you have half a gallon of ice cream and you want to give half of it to a friend, you can find out what part of the ice cream you will give away by multiplying. Your friend will get $\frac{1}{2}$ of $\frac{1}{2}$. When working with fractions, *of* means to multiply.

$$\frac{1}{2} \times \frac{1}{2} = \frac{1}{4}$$

Use These Steps

Find $\frac{1}{2}$ of $\frac{3}{4}$

1. Set up the problem.

   $\frac{1}{2} \times \frac{3}{4} =$

2. Multiply the numerators.

   $\frac{1}{2} \times \frac{3}{4} = \frac{3}{}$

3. Multiply the denominators.

   $\frac{1}{2} \times \frac{3}{4} = \frac{3}{8}$

Multiply the fractions and shade in the figures given to show the answers.

1. $\frac{1}{2}$ of $\frac{1}{3} = \frac{1}{2} \times \frac{1}{3} = \frac{1}{6}$

2. $\frac{1}{2}$ of $\frac{1}{4} =$

3. $\frac{3}{4}$ of $\frac{1}{2} =$

4. $\frac{2}{3}$ of $\frac{1}{3} =$

5. $\frac{3}{5}$ of $\frac{1}{2} =$

6. $\frac{1}{3}$ of $\frac{3}{4} =$

7. $\frac{3}{4} \times \frac{1}{4} =$

8. $\frac{7}{8} \times \frac{1}{3} =$

9. $\frac{1}{5} \times \frac{1}{2} =$

Multiplying Fractions by Fractions

To multiply fractions, set up the problem horizontally. Multiply the numerators and then multiply the denominators. Reduce if possible.

Use These Steps

Multiply $\frac{1}{3} \times \frac{3}{5}$

1. Multiply the numerators.

    $\frac{1}{3} \times \frac{3}{5} = \frac{3}{-}$

2. Multiply the denominators.

    $\frac{1}{3} \times \frac{3}{5} = \frac{3}{15}$

3. Reduce the answer to lowest terms.

    $\frac{3}{15} = \frac{1}{5}$

Multiply. Reduce if possible.

1. $\frac{1}{4} \times \frac{2}{5} = \frac{2}{20} = \frac{1}{10}$

2. $\frac{1}{2} \times \frac{5}{6} =$

3. $\frac{1}{3} \times \frac{3}{7} =$

4. $\frac{1}{8} \times \frac{1}{2} =$

5. $\frac{3}{4} \times \frac{2}{3} =$

6. $\frac{2}{3} \times \frac{2}{7} =$

7. $\frac{3}{5} \times \frac{3}{4} =$

8. $\frac{2}{9} \times \frac{2}{9} =$

9. $\frac{3}{7} \times \frac{4}{7} =$

10. $\frac{5}{6} \times \frac{1}{4} =$

11. $\frac{2}{5} \times \frac{2}{5} =$

12. $\frac{5}{8} \times \frac{1}{10} =$

13. $\frac{2}{9} \times \frac{5}{6} =$

14. $\frac{1}{3} \times \frac{1}{3} =$

15. $\frac{4}{7} \times \frac{1}{2} =$

16. $\frac{3}{4} \times \frac{7}{8} =$

17. $\frac{1}{6} \times \frac{5}{6} =$

18. $\frac{2}{5} \times \frac{5}{8} =$

19. $\frac{6}{7} \times \frac{5}{7} =$

20. $\frac{2}{3} \times \frac{3}{10} =$

21. $\frac{3}{4} \times \frac{3}{4} =$

22. $\frac{1}{7} \times \frac{1}{6} =$

23. $\frac{5}{6} \times \frac{4}{5} =$

24. $\frac{5}{8} \times \frac{3}{8} =$

25. Alex bought $\frac{3}{4}$ pound of potato salad. He and his friends ate $\frac{1}{2}$ of it while watching a football game. How much potato salad did they eat?

26. Daniel drinks $\frac{1}{2}$ quart of milk a day. His younger sister only drinks $\frac{1}{3}$ as much as her brother. How much milk does Daniel's sister drink a day?

Answer_____

Answer_____

Real-Life Application — On the Job

This table shows the average number of cassettes, CDs, and albums sold each day at River City Records. For example, 15 CDs are $\frac{3}{20}$ of total sales.

River City Records

Item	Number Sold	Fraction of Total Sales
Cassettes	40	$\frac{2}{5}$
CDs	15	$\frac{3}{20}$
Albums	45	$\frac{9}{20}$
Total Items Sold	100	

Example $\frac{1}{2}$ of all the cassettes sold are country and western music. What fraction of the total sales is this?

Cassettes are $\frac{2}{5}$ of the total sales. Country and western music is $\frac{1}{2}$ of the cassettes.

$$\frac{1}{2} \times \frac{2}{5} = \frac{2}{10} = \frac{1}{5}$$

$\frac{1}{5}$ of total sales are country and western cassettes.

Solve. Reduce if possible.

1. $\frac{1}{4}$ of the cassettes sold are rap. What fraction of the total sales is this?

   Answer_____

2. $\frac{1}{8}$ of the cassettes sold are classical music. What fraction of the total sales is this?

   Answer_____

3. $\frac{1}{5}$ of the CDs sold are classical music. What fraction of the total sales is this?

   Answer_____

4. $\frac{3}{5}$ of the CDs sold are rock music. What fraction of the total sales is this?

   Answer_____

5. $\frac{1}{3}$ of the albums sold are country and western. What fraction of the total sales is this?

   Answer_____

6. $\frac{4}{9}$ of the albums sold are rock music. What fraction of the total sales is this?

   Answer_____

Multiplying Three Fractions

When multiplying three fractions, multiply the first two numerators. Then multiply that answer by the last numerator. Multiply the denominators the same way.

Use These Steps

Multiply $\frac{1}{2} \times \frac{2}{3} \times \frac{3}{4}$

1. Multiply the first two numerators. Then multiply that answer by the last numerator.

$$\frac{1}{2} \times \frac{2}{3} \times \frac{3}{4} = \frac{6}{}$$
(first two numerators: 2)

2. Multiply the first two denominators. Then multiply that answer by the last denominator.

$$\frac{1}{2} \times \frac{2}{3} \times \frac{3}{4} = \frac{6}{24}$$
(first two denominators: 6)

3. Reduce the answer to lowest terms.

$$\frac{6}{24} = \frac{1}{4}$$

Multiply. Reduce if possible.

1. $\frac{1}{3} \times \frac{2}{3} \times \frac{1}{4} = \frac{2}{36} = \frac{1}{18}$

2. $\frac{1}{2} \times \frac{2}{5} \times \frac{1}{3} =$

3. $\frac{3}{4} \times \frac{3}{4} \times \frac{2}{3} =$

4. $\frac{1}{6} \times \frac{2}{5} \times \frac{1}{2} =$

5. $\frac{2}{3} \times \frac{3}{7} \times \frac{2}{3} =$

6. $\frac{3}{8} \times \frac{1}{2} \times \frac{5}{6} =$

7. $\frac{1}{4} \times \frac{2}{5} \times \frac{3}{5} =$

8. $\frac{2}{9} \times \frac{2}{3} \times \frac{1}{4} =$

9. $\frac{4}{5} \times \frac{5}{6} \times \frac{2}{3} =$

10. $\frac{4}{9} \times \frac{1}{2} \times \frac{3}{5} =$

11. $\frac{3}{4} \times \frac{2}{7} \times \frac{2}{3} =$

12. $\frac{1}{4} \times \frac{3}{5} \times \frac{5}{6} =$

13. $\frac{3}{7} \times \frac{4}{7} \times \frac{1}{2} =$

14. $\frac{2}{5} \times \frac{7}{10} \times \frac{1}{2} =$

15. $\frac{3}{10} \times \frac{2}{5} \times \frac{1}{4} =$

16. $\frac{5}{8} \times \frac{3}{4} \times \frac{2}{3} =$

17. $\frac{1}{5} \times \frac{3}{7} \times \frac{7}{9} =$

18. $\frac{3}{20} \times \frac{2}{3} \times \frac{1}{3} =$

19. $\frac{5}{12} \times \frac{1}{2} \times \frac{2}{5} =$

20. $\frac{3}{4} \times \frac{1}{6} \times \frac{12}{13} =$

21. $\frac{1}{8} \times \frac{4}{5} \times \frac{15}{16} =$

Multiplying Fractions Using Cancellation

You can sometimes make a multiplication problem easier if you cancel before you multiply. To cancel, divide a numerator and a denominator by the same number.

Use These Steps

Multiply $\frac{4}{5} \times \frac{1}{2}$

1. Cancel by dividing the numerator of the first fraction, 4, and the denominator of the second fraction, 2, by 2. Cross out the 4 and the 2 and write the new numbers.

$$\frac{\cancel{4}^{2}}{5} \times \frac{1}{\cancel{2}_{1}} =$$

2. Multiply the numerators of the new fractions.

$$\frac{\cancel{4}^{2}}{5} \times \frac{1}{\cancel{2}_{1}} = \frac{2}{}$$

3. Multiply the denominators of the new fractions.

$$\frac{\cancel{4}^{2}}{5} \times \frac{1}{\cancel{2}_{1}} = \frac{2}{5}$$

Cancel and multiply. Reduce if possible.

1. $\frac{1}{\cancel{9}_{3}} \times \frac{\cancel{3}^{1}}{5} = \frac{1}{15}$

2. $\frac{3}{8} \times \frac{4}{7} =$

3. $\frac{5}{7} \times \frac{3}{5} =$

4. $\frac{2}{3} \times \frac{6}{7} =$

5. $\frac{3}{4} \times \frac{6}{7} =$

6. $\frac{3}{14} \times \frac{4}{7} =$

7. $\frac{3}{8} \times \frac{4}{9} =$

8. $\frac{4}{15} \times \frac{10}{13} =$

9. $\frac{3}{20} \times \frac{5}{8} =$

10. $\frac{1}{6} \times \frac{12}{17} =$

11. $\frac{4}{15} \times \frac{10}{21} =$

12. $\frac{21}{25} \times \frac{9}{14} =$

13. $\frac{1}{10} \times \frac{25}{27} =$

14. $\frac{3}{16} \times \frac{4}{7} =$

15. $\frac{14}{15} \times \frac{11}{35} =$

16. $\frac{6}{11} \times \frac{2}{9} =$

17. $\frac{7}{50} \times \frac{10}{11} =$

18. $\frac{9}{26} \times \frac{13}{14} =$

19. $\frac{8}{9} \times \frac{1}{32} =$

20. $\frac{14}{15} \times \frac{4}{49} =$

21. $\frac{13}{15} \times \frac{4}{39} =$

22. $\frac{3}{25} \times \frac{5}{7} =$

23. $\frac{9}{10} \times \frac{1}{3} =$

24. $\frac{7}{8} \times \frac{16}{19} =$

Multiplying Fractions Using Cancellation

Sometimes you can cancel more than one time when you are multiplying fractions.

Use These Steps

Multiply $\frac{5}{6} \times \frac{3}{10}$

1. Cancel once by dividing the numerator of the first fraction, 5, and the denominator of the second fraction, 10, by 5. Cross out the 5 and the 10. Write the new numbers.

$$\frac{\cancel{5}^{1}}{6} \times \frac{3}{\cancel{10}_{2}} =$$

2. Cancel again by dividing the numerator of the second fraction, 3, and the denominator of the first fraction, 6, by 3. Cross out the 3 and the 6. Write the new numbers.

$$\frac{\cancel{5}^{1}}{\cancel{6}_{2}} \times \frac{\cancel{3}^{1}}{\cancel{10}_{2}} =$$

3. Multiply the new numerators and the new denominators.

$$\frac{\cancel{5}^{1}}{\cancel{6}_{2}} \times \frac{\cancel{3}^{1}}{\cancel{10}_{2}} = \frac{1}{4}$$

Cancel and multiply. Reduce if possible.

1. $\frac{\cancel{9}^{3}}{\cancel{12}_{6}} \times \frac{\cancel{2}^{1}}{\cancel{3}_{1}} = \frac{3}{6} = \frac{1}{2}$

2. $\frac{4}{5} \times \frac{10}{16} =$

3. $\frac{3}{4} \times \frac{12}{15} =$

4. $\frac{3}{6} \times \frac{18}{21} =$

5. $\frac{15}{16} \times \frac{14}{15} =$

6. $\frac{7}{9} \times \frac{6}{21} =$

7. $\frac{9}{10} \times \frac{12}{27} =$

8. $\frac{7}{8} \times \frac{24}{28} =$

9. $\frac{4}{9} \times \frac{3}{14} =$

10. $\frac{22}{24} \times \frac{8}{11} =$

11. $\frac{17}{18} \times \frac{12}{34} =$

12. $\frac{16}{27} \times \frac{9}{20} =$

13. $\frac{5}{12} \times \frac{4}{5} =$

14. $\frac{10}{21} \times \frac{7}{20} =$

15. $\frac{33}{42} \times \frac{6}{22} =$

16. $\frac{7}{24} \times \frac{12}{35} =$

17. The mechanic used a bolt $\frac{15}{16}$ inch long to put in a new water pump. He needed another bolt $\frac{2}{3}$ as long to put on a license plate. How long was the bolt he used for the license plate?

    Answer_____

18. Kenny can mow his back yard in $\frac{3}{4}$ hour. He can mow the front yard in $\frac{2}{3}$ that time. How long does it take Kenny to mow his front yard?

    Answer_____

Multiplying Three Fractions Using Cancellation

When multiplying three fractions, you may be able to cancel several times. Be sure you have canceled all possible numbers before you start to multiply.

Use These Steps

Multiply $\frac{5}{6} \times \frac{3}{4} \times \frac{2}{5}$

1. Start with the numerator of the first fraction, 5. Cancel the numerator of the first fraction with the denominator of the third fraction, 5.

$$\frac{\cancel{5}^{1}}{6} \times \frac{3}{4} \times \frac{2}{\cancel{5}_{1}} =$$

2. Cancel the denominator of the first fraction, 6, with the numerator of the second fraction, 3.

$$\frac{\cancel{5}^{1}}{\cancel{6}_{2}} \times \frac{\cancel{3}^{1}}{4} \times \frac{2}{\cancel{5}_{1}} =$$

3. Cancel the numerator of the third fraction, 2, with the denominator of the second fraction, 4. Multiply the new fractions.

$$\frac{\cancel{5}^{1}}{\cancel{6}_{2}} \times \frac{\cancel{3}^{1}}{\cancel{4}_{2}} \times \frac{\cancel{2}^{1}}{\cancel{5}_{1}} = \frac{1}{4}$$

Cancel and multiply. Reduce if possible.

1. $\dfrac{\cancel{4}^{2}}{\cancel{25}_{5}} \times \dfrac{\cancel{7}^{1}}{\cancel{10}_{5}} \times \dfrac{\cancel{20}^{4}}{\cancel{21}_{3}} = \dfrac{8}{75}$

2. $\dfrac{21}{24} \times \dfrac{8}{35} \times \dfrac{5}{9} =$

3. $\dfrac{8}{10} \times \dfrac{26}{32} \times \dfrac{12}{13} =$

4. $\dfrac{11}{45} \times \dfrac{18}{33} \times \dfrac{1}{2} =$

5. $\dfrac{6}{12} \times \dfrac{10}{30} \times \dfrac{5}{14} =$

6. $\dfrac{15}{36} \times \dfrac{12}{15} \times \dfrac{5}{6} =$

7. $\dfrac{9}{10} \times \dfrac{8}{9} \times \dfrac{3}{4} =$

8. $\dfrac{7}{16} \times \dfrac{6}{21} \times \dfrac{3}{14} =$

9. $\dfrac{5}{8} \times \dfrac{4}{27} \times \dfrac{3}{50} =$

10. $\dfrac{2}{3} \times \dfrac{15}{16} \times \dfrac{4}{5} =$

11. $\dfrac{13}{20} \times \dfrac{5}{39} \times \dfrac{8}{9} =$

12. $\dfrac{24}{35} \times \dfrac{7}{8} \times \dfrac{20}{21} =$

13. $\dfrac{1}{2} \times \dfrac{6}{7} \times \dfrac{7}{12} =$

14. $\dfrac{4}{5} \times \dfrac{25}{32} \times \dfrac{2}{15} =$

15. $\dfrac{17}{50} \times \dfrac{10}{51} \times \dfrac{2}{3} =$

16. $\dfrac{11}{12} \times \dfrac{36}{55} \times \dfrac{5}{9} =$

17. $\dfrac{3}{100} \times \dfrac{75}{81} \times \dfrac{9}{10} =$

18. $\dfrac{7}{15} \times \dfrac{45}{49} \times \dfrac{2}{5} =$

Mixed Review

Add, subtract, or multiply. Cancel and reduce if possible.

1. $\dfrac{3}{4} \times \dfrac{1}{2} =$
2. $\dfrac{2}{3} - \dfrac{1}{3} =$
3. $\dfrac{5}{9} + \dfrac{4}{9} =$
4. $\dfrac{1}{8} \times \dfrac{1}{6} =$

5. $\dfrac{3}{5} \times \dfrac{5}{6} =$
6. $\dfrac{5}{8} - \dfrac{3}{8} =$
7. $\dfrac{5}{7} + \dfrac{3}{7} =$
8. $\dfrac{1}{4} \times \dfrac{3}{8} =$

9. $\dfrac{3}{10} + \dfrac{2}{10} =$
10. $\dfrac{1}{2} \times \dfrac{1}{2} =$
11. $\dfrac{2}{3} \times \dfrac{4}{7} =$
12. $\dfrac{7}{12} - \dfrac{3}{12} =$

13. $\dfrac{5}{6} \times \dfrac{12}{13} =$
14. $\dfrac{3}{7} \times \dfrac{4}{6} =$
15. $\dfrac{4}{10} - \dfrac{2}{10} =$
16. $\dfrac{4}{21} \times \dfrac{7}{9} =$

17. $\dfrac{7}{16} \times \dfrac{4}{14} =$
18. $\dfrac{4}{20} + \dfrac{6}{20} =$
19. $\dfrac{19}{25} - \dfrac{4}{25} =$
20. $\dfrac{7}{30} + \dfrac{1}{30} =$

21. $\dfrac{2}{9} \times \dfrac{3}{8} =$
22. $\dfrac{11}{15} \times \dfrac{5}{22} =$
23. $\dfrac{7}{12} \times \dfrac{2}{21} =$
24. $\dfrac{3}{20} \times \dfrac{5}{9} =$

25. $\dfrac{1}{2} \times \dfrac{4}{5} \times \dfrac{2}{3} =$
26. $\dfrac{3}{8} \times \dfrac{4}{7} \times \dfrac{1}{2} =$
27. $\dfrac{1}{3} \times \dfrac{1}{4} \times \dfrac{1}{5} =$

28. $\dfrac{3}{10} \times \dfrac{4}{9} \times \dfrac{1}{8} =$
29. $\dfrac{16}{20} \times \dfrac{10}{15} \times \dfrac{5}{8} =$
30. $\dfrac{1}{7} \times \dfrac{21}{25} \times \dfrac{10}{12} =$

31. $\dfrac{2}{30} \times \dfrac{6}{10} \times \dfrac{5}{18} =$
32. $\dfrac{12}{13} \times \dfrac{26}{30} \times \dfrac{5}{8} =$
33. $\dfrac{20}{24} \times \dfrac{6}{7} \times \dfrac{4}{5} =$

34. Aurelio had $\dfrac{13}{16}$ pound of potting soil. He used $\dfrac{7}{16}$ pound to repot an African violet. How much potting soil did he have left?

35. Rachel needs to reduce the amount of salt she uses by $\dfrac{1}{2}$. If she used to make soup with $\dfrac{2}{3}$ teaspoon salt, how much salt would she use now?

Answer_____   Answer_____

Multiplying Whole Numbers and Fractions

Multiplying a whole number by a fraction is one way to find part of a whole. For example, to find one half of a dozen, multiply $\frac{1}{2}$ by 12. Remember, when working with fractions, *of* means to multiply.

$$\frac{1}{2} \times 12 = \frac{1}{\cancel{2}} \times \frac{\cancel{12}^{6}}{1} = \frac{6}{1} = 6$$

1 dozen = 12

Use These Steps

Find $\frac{1}{3}$ of a dozen.

1. Set up the problem. Write 12 as an improper fraction with a denominator of 1.

   $\frac{1}{3} \times \frac{12}{1} =$

2. Cancel. Multiply the new fractions.

   $\frac{1}{\cancel{3}} \times \frac{\cancel{12}^{4}}{1} = \frac{4}{1}$

3. The answer is an improper fraction. Change to a whole number by dividing the numerator by the denominator.

   $\frac{4}{1} = 4$

Set up each problem. Cancel and multiply. Use the information in the boxes to help you.

| 1 yard = 36 inches | 1 day = 24 hours | 1 week = 7 days | 1 year = 52 weeks |

1. $\frac{1}{2}$ of a yard = __18__ inches

   $\frac{1}{2} \times 36 = \frac{1}{\cancel{2}} \times \frac{\cancel{36}^{18}}{1} = \frac{18}{1} = 18$

2. $\frac{1}{3}$ of a yard = _____ inches

3. $\frac{3}{4}$ of a yard = _____ inches

4. $\frac{2}{3}$ of a yard = _____ inches

5. $\frac{1}{2}$ of a day = _____ hours

6. $\frac{1}{3}$ of a day = _____ hours

7. $\frac{3}{4}$ of a day = _____ hours

8. $\frac{6}{7}$ of a week = _____ days

9. $\frac{1}{4}$ of a year = _____ weeks

10. $\frac{3}{4}$ of a year = _____ weeks

Multiplying Whole Numbers and Fractions

To multiply a whole number and a fraction, first write the whole number over 1. Then multiply the numerators and the denominators. Remember to change answers that are improper fractions to whole or mixed numbers.

Use These Steps

Multiply $\frac{1}{3} \times 13$

1. Set up the problem. Write 13 as $\frac{13}{1}$.

    $\frac{1}{3} \times \frac{13}{1} =$

2. Multiply the numerators. Multiply the denominators.

    $\frac{1}{3} \times \frac{13}{1} = \frac{13}{3}$

3. The answer is an improper fraction. Change to a mixed number by dividing the numerator by the denominator.

    $\frac{13}{3} = 4\frac{1}{3}$

Multiply. Cancel and reduce if possible.

1. $\frac{2}{3} \times 15 =$

    $\frac{2}{\cancel{3}} \times \frac{\cancel{15}}{1} = \frac{10}{1} = 10$

2. $10 \times \frac{1}{2} =$

3. $8 \times \frac{3}{4} =$

4. $12 \times \frac{1}{4} =$

5. $\frac{2}{7} \times 14 =$

6. $\frac{3}{8} \times 16 =$

7. $\frac{1}{6} \times 18 =$

8. $\frac{4}{5} \times 20 =$

9. $11 \times \frac{1}{2} =$

10. $17 \times \frac{2}{9} =$

11. $\frac{5}{8} \times 15 =$

12. $\frac{2}{3} \times 10 =$

13. $\frac{3}{14} \times 6 =$

14. $18 \times \frac{9}{10} =$

15. $20 \times \frac{5}{6} =$

16. $19 \times \frac{1}{2} =$

17. $\frac{3}{10} \times 20 =$

18. $\frac{2}{7} \times 21 =$

19. $\frac{1}{4} \times 2 =$

20. $\frac{1}{6} \times 5 =$

Multiplying Mixed Numbers and Whole Numbers

When multiplying mixed and whole numbers, remember to change both numbers to improper fractions.

Use These Steps

Multiply $3\frac{1}{6} \times 4$

1. Write $3\frac{1}{6}$ as an improper fraction.

   $3\frac{1}{6} = \frac{19}{6}$

2. Write 4 as an improper fraction.

   $4 = \frac{4}{1}$

3. Cancel. Multiply. Change the answer to a mixed number.

   $\frac{19}{\cancel{6}_3} \times \frac{\cancel{4}^2}{1} = \frac{38}{3} = 12\frac{2}{3}$

Multiply. Cancel and reduce if possible.

1. $2\frac{1}{3} \times 4 =$
   $\frac{7}{3} \times \frac{4}{1} = \frac{28}{3} = 9\frac{1}{3}$

2. $4\frac{2}{5} \times 5 =$

3. $1\frac{1}{8} \times 10 =$

4. $2\frac{5}{9} \times 9 =$

5. $6\frac{2}{3} \times 3 =$

6. $5\frac{1}{2} \times 10 =$

7. $2\frac{3}{7} \times 21 =$

8. $9\frac{3}{4} \times 2 =$

9. $5 \times 3\frac{1}{2} =$

10. $4 \times 7\frac{5}{6} =$

11. $8 \times 3\frac{1}{10} =$

12. $6 \times 6\frac{2}{3} =$

13. $2 \times 1\frac{7}{10} =$

14. $4 \times 3\frac{7}{12} =$

15. $3 \times 5\frac{1}{4} =$

16. $2 \times 9\frac{4}{5} =$

17. $3\frac{6}{7} \times 14 =$

18. $12 \times 1\frac{3}{4} =$

19. $12 \times 2\frac{9}{10} =$

20. $7\frac{1}{7} \times 7 =$

21. Lupe processed 10 catalog orders in one hour. How many orders can she process in $2\frac{1}{2}$ hours?

    Answer _____

22. Jacob ran $2\frac{7}{10}$ miles each day for 5 days. How many miles did he run all together?

    Answer _____

Problem Solving: Figuring Overtime

When you are paid at an hourly rate for regular time and at time and a half for overtime, you can use multiplication to figure your total pay. *Time and a half* means that you get paid your regular rate plus one half your regular rate.

Employee	Regular	Overtime
Kenichi	$ 6	$ 9
Marina	$ 8	
Nestor	$ 12	
Rita	$ 10	
Edward	$ 16	

Example Employees at Marcy's Restaurant are paid time and a half for overtime hours. Overtime hours are hours worked over 40 hours per week.

What is Kenichi's overtime pay rate?

▶ **Step 1.** Find Kenichi's regular pay in the chart.

$6 per hour

▶ **Step 2.** Multiply by $1\frac{1}{2}$ to find his overtime rate.

$$6 \times 1\frac{1}{2} = \frac{\cancel{6}^{3}}{1} \times \frac{3}{\cancel{2}_{1}} = \frac{9}{1} = 9$$

Kenichi's overtime pay rate is $9 per hour.

Solve. Fill in the chart.

1. Find Marina's overtime pay rate.

   Answer_____

2. Find Nestor's overtime pay rate.

   Answer_____

3. Find Rita's overtime pay rate.

   Answer_____

4. Find Edward's overtime pay rate.

   Answer_____

MILLER TIRE FACTORY

EMPLOYEE	OVERTIME RATE	OVERTIME HOURS	OVERTIME PAY
Oscar	$ 6	$3\frac{1}{2}$	
Carl	$ 9	$5\frac{2}{3}$	
Ruthie	$ 15	$2\frac{1}{3}$	
James	$ 18	$8\frac{1}{2}$	
Mike	$ 12	$5\frac{3}{4}$	$ 69

Example Employees at Miller Tire Factory are paid by the hour. When they work over 40 hours a week, they get time and a half for those hours.

Mike worked $5\frac{3}{4}$ hours overtime this week. How much will he earn in overtime pay?

▶ **Step 1.** Find Mike's overtime pay rate on the chart.

$12

▶ **Step 2.** Multiply to find Mike's overtime pay.

$$12 \times 5\frac{3}{4} = \frac{\cancel{12}^{3}}{1} \times \frac{23}{\cancel{4}_{1}} = \frac{69}{1} = 69$$

Mike will get $69 in overtime pay.

Solve. Fill in the chart.

1. Oscar worked $3\frac{1}{2}$ hours overtime last week. What was his overtime pay?

   Answer_____

2. Carl worked $5\frac{2}{3}$ hours overtime. What was his overtime pay?

   Answer_____

3. Ruthie worked $2\frac{1}{3}$ hours overtime. What was her overtime pay?

   Answer_____

4. James worked $8\frac{1}{2}$ hours overtime. What was his overtime pay?

   Answer_____

Multiplying Mixed Numbers and Fractions

When multiplying mixed numbers and fractions, change the mixed number to an improper fraction. Cancel if possible. Change answers that are improper fractions to mixed numbers.

Use These Steps

Multiply $6\frac{3}{7} \times \frac{3}{5}$

1. Write $6\frac{3}{7}$ as an improper fraction.

    $6\frac{3}{7} = \frac{45}{7}$

2. Cancel. Multiply the numerators. Multiply the denominators.

    $\frac{\cancel{45}^{9}}{7} \times \frac{3}{\cancel{5}_{1}} = \frac{27}{7}$

3. Change the answer to a mixed number.

    $\frac{27}{7} = 3\frac{6}{7}$

Multiply. Cancel and reduce if possible.

1. $3\frac{1}{5} \times \frac{3}{4} =$
   $\frac{\cancel{16}^{4}}{5} \times \frac{3}{\cancel{4}_{1}} = \frac{12}{5} = 2\frac{2}{5}$

2. $4\frac{3}{8} \times \frac{5}{7} =$

3. $5\frac{2}{5} \times \frac{1}{4} =$

4. $7\frac{1}{9} \times \frac{3}{4} =$

5. $2\frac{1}{2} \times \frac{3}{10} =$

6. $1\frac{1}{3} \times \frac{3}{16} =$

7. $6\frac{1}{3} \times \frac{3}{5} =$

8. $3\frac{2}{7} \times \frac{1}{7} =$

9. $\frac{3}{8} \times 4\frac{1}{6} =$

10. $3\frac{7}{9} \times \frac{3}{4} =$

11. $\frac{9}{10} \times 5\frac{2}{3} =$

12. $4\frac{1}{4} \times \frac{12}{13} =$

13. $\frac{1}{2} \times 9\frac{2}{3} =$

14. $\frac{7}{8} \times 3\frac{1}{14} =$

15. $\frac{5}{12} \times 10\frac{1}{10} =$

16. $\frac{7}{15} \times 4\frac{3}{21} =$

17. $6\frac{1}{3} \times \frac{2}{9} =$

18. $\frac{8}{11} \times 2\frac{1}{2} =$

Multiplying Mixed Numbers and Mixed Numbers

When multiplying mixed numbers, change both mixed numbers to improper fractions. Cancel if possible. Multiply. Change answers that are improper fractions to whole or mixed numbers.

Use These Steps

Multiply $2\frac{1}{7} \times 1\frac{2}{5}$

1. Write $2\frac{1}{7}$ as an improper fraction.

   $2\frac{1}{7} = \frac{15}{7}$

2. Write $1\frac{2}{5}$ as an improper fraction.

   $1\frac{2}{5} = \frac{7}{5}$

3. Cancel. Multiply. Change the answer to a whole number.

   $\dfrac{\cancel{15}^{3}}{\cancel{7}_{1}} \times \dfrac{\cancel{7}^{1}}{\cancel{5}_{1}} = \dfrac{3}{1} = 3$

Multiply. Cancel and reduce if possible.

1. $3\frac{1}{9} \times 5\frac{1}{4} =$
   $\dfrac{\cancel{28}}{\cancel{9}_{3}}^{7} \times \dfrac{\cancel{21}^{7}}{\cancel{4}_{1}} = \dfrac{49}{3} = 16\frac{1}{3}$

2. $9\frac{4}{5} \times 3\frac{4}{7} =$

3. $4\frac{7}{8} \times 9\frac{1}{3} =$

4. $2\frac{3}{4} \times 2\frac{5}{11} =$

5. $4\frac{1}{5} \times 3\frac{1}{3} =$

6. $1\frac{1}{5} \times 2\frac{5}{6} =$

7. $5\frac{1}{3} \times 6\frac{1}{8} =$

8. $1\frac{1}{2} \times 7\frac{1}{3} =$

9. $3\frac{1}{10} \times 4\frac{2}{7} =$

10. $1\frac{5}{9} \times 2\frac{9}{14} =$

11. $5\frac{1}{7} \times 2\frac{1}{10} =$

12. $1\frac{1}{12} \times 3\frac{3}{5} =$

13. $7\frac{1}{8} \times 6\frac{2}{3} =$

14. $5\frac{1}{11} \times 4\frac{1}{2} =$

15. $9\frac{3}{8} \times 8\frac{8}{25} =$

Real-Life Application

Time Off

Robert has a recipe that makes 1 dozen large brownies. He needs to make $3\frac{1}{2}$ dozen brownies for his school's bake sale. He will need to use $3\frac{1}{2}$ times as much of each ingredient listed in the recipe.

Here's what's cookin' **Chewy Fudge Brownies**
Recipe from the kitchen of **Robert**
Serves **1 Dozen**

- $\frac{1}{2}$ cup butter   2 eggs
- $1\frac{1}{2}$ ounces unsweetened chocolate
- $1\frac{1}{4}$ cups sugar $\frac{3}{4}$ cup flour
- $\frac{1}{2}$ teaspoon baking powder
- $\frac{1}{8}$ teaspoon salt $\frac{3}{4}$ cup chopped nuts

Example How many cups of butter will Robert need to make $3\frac{1}{2}$ dozen brownies?

$$\frac{1}{2} \text{ cup} \times 3\frac{1}{2} = \frac{1}{2} \times \frac{7}{2} = \frac{7}{4} = 1\frac{3}{4}$$

Robert needs $1\frac{3}{4}$ cups butter.

Solve. Reduce if possible.

1. How many ounces of unsweetened chocolate will Robert need to make $3\frac{1}{2}$ dozen brownies?

    Answer_____

2. How many cups of sugar will Robert need to make $3\frac{1}{2}$ dozen brownies?

    Answer_____

3. How many cups of flour will Robert need to make $3\frac{1}{2}$ dozen brownies?

    Answer_____

4. How many teaspoons of salt will he need to make $3\frac{1}{2}$ dozen brownies?

    Answer_____

5. For the church bake sale, Robert needs to make 5 dozen brownies. How many cups of butter will he need?

    Answer_____

6. How many cups of chopped nuts will he need to make 5 dozen brownies?

    Answer_____

Unit 4 Review

Multiply. Reduce if possible.

1. $\frac{1}{3} \times \frac{2}{3} =$
2. $\frac{1}{2} \times \frac{1}{3} =$
3. $\frac{1}{4} \times \frac{1}{3} =$
4. $\frac{3}{5} \times \frac{1}{2} =$

5. $\frac{5}{9} \times \frac{1}{4} =$
6. $\frac{4}{5} \times \frac{4}{5} =$
7. $\frac{5}{6} \times \frac{7}{8} =$
8. $\frac{7}{10} \times \frac{1}{2} =$

9. $\frac{1}{2} \times \frac{1}{3} \times \frac{1}{4} =$
10. $\frac{3}{5} \times \frac{1}{4} \times \frac{3}{7} =$
11. $\frac{1}{5} \times \frac{3}{4} \times \frac{1}{4} =$

12. $\frac{2}{3} \times \frac{3}{5} \times \frac{5}{8} =$
13. $\frac{1}{10} \times \frac{5}{6} \times \frac{2}{7} =$
14. $\frac{4}{9} \times \frac{1}{3} \times \frac{3}{5} =$

Multiply. Cancel and reduce if possible.

15. $\frac{1}{3} \times \frac{3}{7} =$
16. $\frac{4}{5} \times \frac{10}{12} =$
17. $\frac{3}{8} \times \frac{4}{9} =$
18. $\frac{2}{15} \times \frac{5}{6} =$

19. $\frac{4}{10} \times \frac{3}{8} =$
20. $\frac{2}{7} \times \frac{3}{10} =$
21. $\frac{5}{14} \times \frac{7}{10} =$
22. $\frac{1}{6} \times \frac{6}{7} =$

23. $\frac{1}{4} \times \frac{2}{5} =$
24. $\frac{2}{7} \times \frac{7}{9} =$
25. $\frac{3}{16} \times \frac{8}{9} =$
26. $\frac{1}{8} \times \frac{8}{9} =$

27. $\frac{4}{9} \times \frac{3}{4} \times \frac{3}{5} =$
28. $\frac{8}{10} \times \frac{12}{14} \times \frac{7}{16} =$
29. $\frac{3}{25} \times \frac{5}{12} \times \frac{8}{9} =$

30. $\frac{11}{45} \times \frac{18}{33} \times \frac{1}{2} =$
31. $\frac{6}{12} \times \frac{10}{30} \times \frac{5}{14} =$
32. $\frac{15}{36} \times \frac{12}{15} \times \frac{5}{6} =$

33. $\frac{5}{8} \times \frac{1}{6} \times \frac{9}{10} =$
34. $\frac{4}{11} \times \frac{5}{12} \times \frac{11}{15} =$
35. $\frac{4}{15} \times \frac{7}{12} \times \frac{3}{4} =$

Multiply. Cancel and reduce if possible.

36. $\frac{1}{2} \times 10 =$

37. $\frac{2}{3} \times 12 =$

38. $\frac{5}{6} \times 18 =$

39. $\frac{4}{5} \times 30 =$

40. $5 \times \frac{2}{3} =$

41. $7 \times \frac{1}{4} =$

42. $9 \times \frac{2}{5} =$

43. $11 \times \frac{1}{2} =$

44. $20 \times \frac{2}{5} =$

45. $\frac{4}{9} \times 6 =$

46. $42 \times \frac{1}{6} =$

47. $\frac{5}{14} \times 28 =$

Multiply. Cancel and reduce if possible.

48. $1\frac{1}{2} \times 3 =$

49. $6 \times 5\frac{2}{3} =$

50. $4 \times 2\frac{3}{8} =$

51. $3\frac{6}{7} \times 9 =$

52. $2\frac{1}{3} \times \frac{3}{5} =$

53. $\frac{5}{9} \times 2\frac{1}{3} =$

54. $\frac{7}{15} \times 5\frac{1}{14} =$

55. $4\frac{5}{11} \times \frac{2}{7} =$

56. $1\frac{1}{6} \times 2\frac{5}{7} =$

57. $2\frac{2}{3} \times 5\frac{1}{4} =$

58. $5\frac{1}{7} \times 2\frac{1}{6} =$

59. $1\frac{1}{5} \times 6\frac{2}{3} =$

Below is a list of the problems in this review and the pages on which the skills are taught. If you missed any problems, turn to the pages listed and practice the skills. Then correct the problems you missed in the Unit Review.

Problems	Pages	Problems	Pages
1-8	109-110	36-47	117-118
9-14	112	48-51	119
15-26	113-114	52-55	122
27-35	115	56-59	123

Unit 5 DIVIDING FRACTIONS

Have you ever needed to divide $\frac{1}{2}$ a cake equally among four people, divide a pound of peanuts into $\frac{1}{4}$-pound packages, or cut a 6-foot piece of wood into $1\frac{1}{2}$-foot pieces? All of these problems can be solved by dividing fractions.

You will find that knowing how to multiply fractions will help you learn how to divide fractions.

In this unit, you will learn how to work division problems with fractions, whole numbers, and mixed numbers, and how to solve word problems using fractions.

Getting Ready

You should be familiar with the skills on this page and the next before you begin this unit. To check your answers, turn to page 199.

 Knowing how to multiply fractions is the first step in learning how to divide fractions.

Multiply these fractions by canceling.

1. $\dfrac{1}{\cancel{2}} \times \dfrac{\cancel{4}^{2}}{9} = \dfrac{2}{9}$

2. $\dfrac{2}{3} \times \dfrac{6}{7} =$

3. $\dfrac{5}{8} \times \dfrac{16}{20} =$

4. $\dfrac{3}{4} \times \dfrac{8}{9} =$

5. $\dfrac{1}{3} \times \dfrac{3}{12} =$

6. $\dfrac{3}{8} \times \dfrac{4}{21} =$

7. $\dfrac{11}{16} \times \dfrac{4}{9} =$

8. $\dfrac{7}{11} \times \dfrac{33}{49} =$

9. $\dfrac{4}{15} \times \dfrac{3}{7} =$

10. $\dfrac{5}{6} \times \dfrac{24}{35} =$

For review, see Unit 4, pages 113–114.   127

Getting Ready

 Fractions can also be multiplied by whole or mixed numbers.

Multiply. Cancel and reduce if possible.

11. $\frac{1}{5} \times 30 = 6$
12. $17 \times \frac{1}{4} =$
13. $\frac{2}{3} \times 45 =$
14. $24 \times \frac{1}{6} =$
15. $\frac{3}{7} \times 21 =$

16. $2\frac{1}{5} \times \frac{10}{11} =$
17. $\frac{7}{10} \times 1\frac{1}{7} =$
18. $\frac{8}{9} \times 3\frac{1}{2} =$
19. $7\frac{3}{4} \times \frac{2}{3} =$
20. $4\frac{1}{10} \times \frac{5}{7} =$

For review, see Unit 4, pages 118 and 122.

 When your answer is an improper fraction, you need to change it to a whole or mixed number.

Change each improper fraction to a whole or mixed number. Reduce if possible.

21. $\frac{10}{2} = 5$
22. $\frac{13}{5} =$
23. $\frac{17}{3} =$
24. $\frac{15}{4} =$
25. $\frac{20}{5} =$

26. $\frac{31}{6} =$
27. $\frac{9}{4} =$
28. $\frac{15}{10} =$
29. $\frac{18}{9} =$
30. $\frac{48}{12} =$

For review, see Unit 1, pages 33–34.

 All fractions in your answer should be reduced to lowest terms.

Reduce each fraction to lowest terms. If it is in lowest terms, write LT.

31. $\frac{2}{4} = \frac{1}{2}$
32. $\frac{3}{8} =$
33. $\frac{3}{9} =$
34. $\frac{6}{12} =$
35. $\frac{7}{8} =$

36. $\frac{4}{10} =$
37. $\frac{6}{9} =$
38. $\frac{15}{20} =$
39. $\frac{21}{49} =$
40. $\frac{16}{24} =$

For review, see Unit 1, pages 15–16.

 To divide mixed numbers, you will need to know how to change a mixed number to an improper fraction.

Change each mixed number to an improper fraction.

41. $1\frac{1}{2} = \frac{3}{2}$
42. $2\frac{2}{3} =$
43. $4\frac{1}{4} =$
44. $3\frac{2}{5} =$
45. $5\frac{3}{8} =$

For review, see Unit 1, pages 35–36.

Dividing Fractions by Fractions

To find out how many equal parts are in a fraction, you may need to divide a fraction by a fraction.

For example, if you have $\frac{3}{4}$ yard of material to divide equally into $\frac{1}{4}$-yard pieces, how many pieces will you get? To find how many $\frac{1}{4}$s are in $\frac{3}{4}$s, divide $\frac{3}{4}$ by $\frac{1}{4}$.

To divide by a fraction, invert (turn upside down) the fraction you are dividing by (the fraction to the right of the division sign). When you invert $\frac{1}{4}$, you get $\frac{4}{1}$. Change the division sign to a multiplication sign and multiply.

$$\frac{3}{4} \div \frac{1}{4} = \frac{3}{4} \times \frac{4}{1} = \frac{12}{4} = 3 \text{ pieces}$$

Use These Steps

Divide $\frac{1}{2} \div \frac{1}{8}$

1. Invert the fraction to the right of the division sign.

   $\frac{1}{8} \diagup \frac{8}{1}$

2. Change the division sign to a multiplication sign. Multiply by the new fraction, $\frac{8}{1}$. Cancel.

   $\frac{1}{\underset{1}{2}} \times \frac{\overset{4}{8}}{1} = \frac{4}{1}$

3. Change the answer to a whole number.

   $\frac{4}{1} = 4$

Divide. Cancel and reduce if possible.

1. $\frac{5}{6} \div \frac{1}{6} = \frac{5}{\underset{1}{6}} \times \frac{\overset{1}{6}}{1} = \frac{5}{1} = 5$

2. $\frac{3}{4} \div \frac{1}{8} =$

3. $\frac{2}{5} \div \frac{1}{10} =$

4. $\frac{2}{3} \div \frac{1}{3} =$

5. $\frac{3}{8} \div \frac{1}{16} =$

6. $\frac{7}{10} \div \frac{1}{20} =$

7. $\frac{6}{9} \div \frac{1}{9} =$

8. $\frac{3}{7} \div \frac{1}{21} =$

9. $\frac{5}{9} \div \frac{1}{18} =$

10. $\frac{2}{3} \div \frac{1}{6} =$

Dividing Fractions by Fractions

When you divide fractions, always invert the fraction to the right of the division sign and then multiply.

Use These Steps

Divide $\frac{4}{5} \div \frac{2}{7}$

1. Invert the fraction to the right of the division sign.

$$\frac{2}{7} \searrow \frac{7}{2}$$

2. Multiply by the new fraction. Cancel.

$$\frac{\overset{2}{\cancel{4}}}{5} \times \frac{7}{\underset{1}{\cancel{2}}} = \frac{14}{5}$$

3. Change the answer to a mixed number.

$$\frac{14}{5} = 2\frac{4}{5}$$

Divide. Cancel and reduce if possible.

1. $\frac{3}{4} \div \frac{3}{5} =$

$\frac{\cancel{3}}{4} \times \frac{5}{\underset{1}{\cancel{3}}} = \frac{5}{4} = 1\frac{1}{4}$

2. $\frac{5}{9} \div \frac{1}{4} =$

3. $\frac{4}{7} \div \frac{3}{8} =$

4. $\frac{7}{10} \div \frac{7}{12} =$

5. $\frac{1}{3} \div \frac{2}{3} =$

$\frac{1}{\cancel{3}} \times \frac{\overset{1}{\cancel{3}}}{2} = \frac{1}{2}$

6. $\frac{2}{5} \div \frac{5}{9} =$

7. $\frac{1}{4} \div \frac{1}{3} =$

8. $\frac{3}{7} \div \frac{3}{5} =$

9. $\frac{2}{3} \div \frac{3}{4} =$

10. $\frac{9}{10} \div \frac{2}{5} =$

11. $\frac{5}{8} \div \frac{3}{4} =$

12. $\frac{7}{9} \div \frac{1}{6} =$

13. $\frac{5}{8} \div \frac{2}{5} =$

14. $\frac{4}{5} \div \frac{10}{11} =$

15. $\frac{5}{6} \div \frac{1}{3} =$

16. $\frac{1}{8} \div \frac{1}{4} =$

17. Karen has $\frac{1}{2}$ yard of wire that she wants to cut into $\frac{1}{4}$-yard pieces. How many pieces will she get?

Answer _____

18. If Karen cuts her $\frac{1}{2}$ yard of wire into $\frac{1}{6}$-yard pieces, how many pieces will she get?

Answer _____

Dividing Whole Numbers by Fractions

To divide a whole number by a fraction, change the whole number to an improper fraction with a denominator of 1. Then invert the fraction to the right of the division sign and multiply the new fractions.

For example, if you have 3 acres of land and want to find out how many $\frac{1}{4}$-acre plots you have, divide 3 by $\frac{1}{4}$.

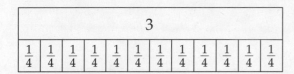

$3 \div \frac{1}{4} = \frac{3}{1} \times \frac{4}{1} = \frac{12}{1} = 12$ plots

Use These Steps

Divide $4 \div \frac{1}{6}$

1. Change the whole number, 4, to an improper fraction with a denominator of 1.

    $4 = \frac{4}{1}$

2. Invert the fraction to the right of the division sign.

    $\frac{1}{6} \searrow \frac{6}{1}$

3. Multiply the new fractions. Change the answer to a whole number.

    $\frac{4}{1} \times \frac{6}{1} = \frac{24}{1} = 24$

Divide. Reduce if possible.

1.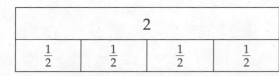

   $2 \div \frac{1}{2} = \frac{2}{1} \times \frac{2}{1} = \frac{4}{1} = 4$

2.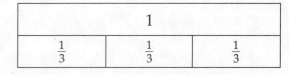

   $1 \div \frac{1}{3} =$

3. $2 \div \frac{1}{4} =$

4. $3 \div \frac{1}{6} =$

5. $4 \div \frac{1}{4} =$

6. $5 \div \frac{1}{5} =$

7. $5 \div \frac{1}{4} =$

8. $8 \div \frac{1}{3} =$

9. $6 \div \frac{1}{9} =$

10. $7 \div \frac{1}{2} =$

11. $15 \div \frac{1}{3} =$

12. $12 \div \frac{1}{4} =$

13. $20 \div \frac{1}{5} =$

14. $5 \div \frac{1}{8} =$

Dividing Whole Numbers by Fractions

Remember to invert only the fraction to the right of the division sign and then multiply.

Use These Steps

Divide $2 \div \frac{3}{4}$

1. Change the whole number, 2, to an improper fraction with a denominator of 1.

   $2 = \frac{2}{1}$

2. Invert the fraction to the right of the division sign.

   $\frac{3}{4} \rightarrow \frac{4}{3}$

3. Multiply the new fractions. Change the answer to a mixed number.

   $\frac{2}{1} \times \frac{4}{3} = \frac{8}{3} = 2\frac{2}{3}$

Divide. Cancel and reduce if possible.

1. $3 \div \frac{2}{5} =$

   $\frac{3}{1} \times \frac{5}{2} = \frac{15}{2} = 7\frac{1}{2}$

2. $5 \div \frac{3}{4} =$

3. $10 \div \frac{2}{3} =$

4. $9 \div \frac{5}{6} =$

5. $12 \div \frac{3}{7} =$

6. $7 \div \frac{4}{9} =$

7. $15 \div \frac{3}{5} =$

8. $13 \div \frac{7}{8} =$

9. $6 \div \frac{3}{10} =$

10. $20 \div \frac{1}{3} =$

11. $18 \div \frac{2}{9} =$

12. $21 \div \frac{7}{11} =$

13. $7 \div \frac{1}{3} =$

14. $6 \div \frac{5}{8} =$

15. $10 \div \frac{1}{5} =$

16. $13 \div \frac{3}{4} =$

17. Jana bought two pieces of plastic sheeting. If she cuts each piece of plastic into twelfths, how many pieces will she get?
    (Hint: Divide 2 by $\frac{1}{12}$.)

18. If Jana cut both pieces of plastic into ninths, how many pieces will she get?

Answer _____

Answer _____

Dividing Mixed Numbers by Fractions

To divide a mixed number by a fraction, first change the mixed number to an improper fraction. Then invert the fraction to the right of the division sign and multiply.

For example, if you have $1\frac{1}{2}$ pounds of ground beef, how many $\frac{1}{4}$-pound hamburgers can you make?

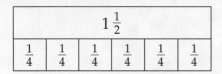

$$1\frac{1}{2} \div \frac{1}{4} = \frac{3}{\cancel{2}_1} \times \frac{\cancel{4}^2}{1} = \frac{6}{1} = 6 \text{ hamburgers}$$

Use These Steps

Divide $2\frac{3}{4} \div \frac{1}{8}$

1. Change the mixed number, $2\frac{3}{4}$, to an improper fraction.

    $2\frac{3}{4} = \frac{11}{4}$

2. Invert the fraction to the right of the division sign.

    $\frac{1}{8} \diagdown \frac{8}{1}$

3. Multiply the new fractions. Cancel. Change the answer to a whole number.

    $\frac{11}{\cancel{4}_1} \times \frac{\cancel{8}^2}{1} = \frac{22}{1} = 22$

Divide. Cancel and reduce if possible.

1.

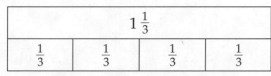

$1\frac{1}{3} \div \frac{1}{3} = \frac{4}{\cancel{3}_1} \times \frac{\cancel{3}^1}{1} = \frac{4}{1} = 4$

2.

$2\frac{1}{2} \div \frac{1}{4} =$

3. $1\frac{2}{5} \div \frac{1}{10} =$

4. $3\frac{3}{4} \div \frac{1}{4} =$

5. $2\frac{5}{8} \div \frac{1}{8} =$

6. $1\frac{2}{3} \div \frac{1}{6} =$

7. $3\frac{2}{7} \div \frac{1}{7} =$

8. $4\frac{1}{4} \div \frac{1}{16} =$

9. $5\frac{4}{9} \div \frac{1}{9} =$

10. $6\frac{3}{10} \div \frac{1}{20} =$

11. $2\frac{2}{5} \div \frac{1}{15} =$

133

Dividing Mixed Numbers by Fractions

When you divide a mixed number by a fraction, the answer may be an improper fraction. Change the improper fraction to a whole number or a mixed number.

Use These Steps

Divide $3\frac{1}{8} \div \frac{1}{4}$

1. Change $3\frac{1}{8}$ to an improper fraction.

    $3\frac{1}{8} = \frac{25}{8}$

2. Invert the fraction to the right of the division sign.

    $\frac{1}{4} \rightarrow \frac{4}{1}$

3. Multiply the new fractions. Cancel. Change the answer to a mixed number.

    $\frac{25}{\cancel{8}_2} \times \frac{\cancel{4}^1}{1} = \frac{25}{2} = 12\frac{1}{2}$

Divide. Cancel and reduce if possible.

1. $4\frac{1}{2} \div \frac{2}{3} =$
   $\frac{9}{2} \times \frac{3}{2} = \frac{27}{4} = 6\frac{3}{4}$

2. $5\frac{2}{5} \div \frac{1}{8} =$

3. $2\frac{7}{8} \div \frac{3}{4} =$

4. $6\frac{1}{9} \div \frac{1}{3} =$

5. $7\frac{5}{7} \div \frac{1}{6} =$

6. $10\frac{4}{5} \div \frac{2}{9} =$

7. $8\frac{3}{4} \div \frac{1}{2} =$

8. $1\frac{5}{8} \div \frac{4}{5} =$

9. $3\frac{6}{11} \div \frac{2}{3} =$

10. $9\frac{1}{14} \div \frac{2}{7} =$

11. $12\frac{5}{6} \div \frac{5}{18} =$

12. $15\frac{7}{20} \div \frac{1}{10} =$

13. $3\frac{1}{3} \div \frac{5}{7} =$

14. $8\frac{1}{4} \div \frac{9}{10} =$

15. $5\frac{2}{9} \div \frac{2}{3} =$

16. $2\frac{5}{6} \div \frac{15}{16} =$

17. Joe prepared $3\frac{1}{2}$ pounds of dough to make bread. He used $\frac{3}{4}$ pound of dough for each loaf of bread. How many loaves of bread did he make?

18. Betty is making pecan cookies. She has $5\frac{1}{3}$ cups of pecans. If she needs $\frac{1}{2}$ cup for each dozen, how many dozen cookies can she make?

Answer _____

Answer _____

Real Life Application — At the Store

The Smithville Food Co-op buys some items in large packages. Gracie's job is to divide the large packages into smaller packages of equal size. Then she puts the packages out for sale.

Example The co-op buys granola in 20-pound bags. If Gracie divides one 20-pound bag into $\frac{3}{4}$-pound bags, how many of the smaller bags can she get? How much granola will be left over?

$$20 \div \frac{3}{4} = \frac{20}{1} \times \frac{4}{3} = \frac{80}{3} = 26\frac{2}{3}$$

Gracie will get 26 whole bags. She will have $\frac{2}{3}$ pound of granola left over.

Solve.

1. The co-op buys cheddar cheese in 10-pound blocks. If Gracie divides one block into $\frac{1}{2}$-pound pieces, how many pieces can she get?

   Answer _____

2. The co-op bought a box of mixed nuts weighing 18 pounds. How many $\frac{1}{4}$-pound packages can Gracie get from the whole box?

   Answer _____

3. The co-op buys some spices in $\frac{1}{2}$-pound packages. If Gracie divides the spices into $\frac{1}{16}$-pound packages, how many smaller packages can she get from each $\frac{1}{2}$-pound package?

   Answer _____

4. The co-op bought a can of pepper weighing 1 pound. How many $\frac{1}{8}$-pound packages can Gracie make from the can?

   Answer _____

5. The co-op buys popcorn in 5-pound bags. One bag weighed only $4\frac{9}{10}$ pounds. How many $\frac{2}{3}$-pound packages can Gracie get from that bag? How much popcorn is left over?

   Answer _____

6. The co-op bought a bushel of corn from a farmer. The corn weighed $25\frac{3}{4}$ pounds. How many $\frac{3}{4}$-pound packages can Gracie get from the bushel? How much corn is left over?

   Answer _____

Mixed Review

Add, subtract, multiply, or divide. Cancel and reduce if possible.

1. $\dfrac{2}{3} \div \dfrac{1}{6} =$
2. $\dfrac{1}{2} + \dfrac{1}{3} =$
3. $\dfrac{2}{5} - \dfrac{1}{4} =$
4. $\dfrac{1}{2} \div \dfrac{1}{8} =$

5. $\dfrac{9}{10} \div \dfrac{3}{5} =$
6. $\dfrac{5}{7} - \dfrac{3}{8} =$
7. $\dfrac{1}{4} \times \dfrac{1}{3} =$
8. $\dfrac{2}{5} \times \dfrac{10}{12} =$

9. $\dfrac{4}{9} - \dfrac{1}{3} =$
10. $\dfrac{4}{10} \div \dfrac{2}{5} =$
11. $\dfrac{7}{12} + \dfrac{3}{10} =$
12. $\dfrac{3}{4} \times \dfrac{1}{2} =$

13. $4 \times \dfrac{2}{3} =$
14. $3 \div \dfrac{1}{2} =$
15. $2 + \dfrac{8}{9} =$
16. $9 \div \dfrac{1}{3} =$

17. $7 - \dfrac{4}{5} =$
18. $6 + \dfrac{3}{7} =$
19. $4 \div \dfrac{5}{8} =$
20. $10 \times \dfrac{5}{6} =$

21. $8 \div \dfrac{1}{4} =$
22. $12 \div \dfrac{7}{10} =$
23. $15 \times \dfrac{11}{30} =$
24. $20 \div \dfrac{5}{12} =$

25. $2\dfrac{5}{8} \div \dfrac{1}{8} =$
26. $1\dfrac{1}{2} \div \dfrac{3}{4} =$
27. $4\dfrac{2}{3} \times \dfrac{1}{3} =$
28. $7\dfrac{6}{10} \div \dfrac{1}{5} =$

29. $5\dfrac{3}{16} \times \dfrac{1}{2} =$
30. $1\dfrac{3}{10} \div \dfrac{4}{5} =$
31. $8\dfrac{2}{9} \div \dfrac{2}{3} =$
32. $12\dfrac{1}{5} \div \dfrac{3}{8} =$

33. $3\dfrac{3}{4} \div \dfrac{5}{16} =$
34. $6\dfrac{6}{7} \times \dfrac{2}{6} =$
35. $20\dfrac{9}{10} \div \dfrac{3}{4} =$
36. $7\dfrac{2}{3} \times \dfrac{5}{6} =$

Dividing Fractions by Whole Numbers

To divide a fraction by a whole number, change the whole number to an improper fraction with a denominator of 1. Invert and multiply.

For example, if you have $\frac{1}{2}$ a cake and want to divide it evenly among 3 people, how much of the cake does each person get?

 $\frac{1}{2} \div 3 = \frac{1}{2} \div \frac{3}{1} = \frac{1}{2} \times \frac{1}{3} = \frac{1}{6}$ of the cake

Use These Steps

Divide $\frac{3}{4} \div 2$

1. Change the whole number, 2, to an improper fraction with a denominator of 1.

    $2 = \frac{2}{1}$

2. Invert the improper fraction, $\frac{2}{1}$.

3. Multiply $\frac{3}{4}$ by the new fraction.

    $\frac{3}{4} \times \frac{1}{2} = \frac{3}{8}$

Divide. Cancel and reduce if possible.

1.

    $\frac{1}{4} \div 2 = \frac{1}{4} \div \frac{2}{1} = \frac{1}{4} \times \frac{1}{2} = \frac{1}{8}$

2.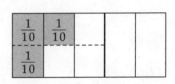

    $\frac{3}{5} \div 2 =$

3. $\frac{2}{3} \div 4 =$

4. $\frac{1}{2} \div 5 =$

5. $\frac{1}{7} \div 2 =$

6. $\frac{1}{8} \div 6 =$

7. $\frac{2}{5} \div 4 =$

8. $\frac{2}{9} \div 7 =$

9. $\frac{3}{8} \div 3 =$

10. $\frac{7}{10} \div 3 =$

11. $\frac{8}{9} \div 16 =$

12. $\frac{5}{6} \div 3 =$

13. $\frac{4}{5} \div 5 =$

14. $\frac{1}{3} \div 12 =$

Dividing Fractions by Whole Numbers

Remember, first change the whole number to an improper fraction. Then invert and multiply. Reduce to lowest terms if possible.

Use These Steps

Divide $\frac{5}{7} \div 9$

1. Change the whole number, 9, to an improper fraction with a denominator of 1.

    $9 = \frac{9}{1}$

2. Invert the improper fraction, $\frac{9}{1}$.

    $\frac{9}{1} \rightarrow \frac{1}{9}$

3. Multiply $\frac{5}{7}$ by the new fraction.

    $\frac{5}{7} \times \frac{1}{9} = \frac{5}{63}$

Divide. Cancel and reduce if possible.

1. $\frac{2}{3} \div 2 =$

    $\frac{2}{3} \div \frac{2}{1} = \frac{\cancel{2}^1}{3} \times \frac{1}{\cancel{2}_1} = \frac{1}{3}$

2. $\frac{3}{4} \div 6 =$

3. $\frac{7}{8} \div 5 =$

4. $\frac{1}{6} \div 3 =$

5. $\frac{6}{7} \div 6 =$

6. $\frac{9}{10} \div 4 =$

7. $\frac{5}{12} \div 10 =$

8. $\frac{15}{16} \div 5 =$

9. $\frac{7}{11} \div 7 =$

10. $\frac{3}{10} \div 6 =$

11. $\frac{5}{14} \div 15 =$

12. $\frac{1}{8} \div 3 =$

13. Maxine has $\frac{3}{4}$ pound of mocha coffee she wants to split with her neighbor. If she divides the coffee evenly with her neighbor, how much coffee will each one get?

14. Marie has $\frac{7}{8}$ yard of fabric. She wants to make 3 scarves. How much material can she use for each scarf?

Answer _____

Answer _____

Dividing Mixed Numbers by Whole Numbers

To divide a mixed number by a whole number, change both the mixed number and the whole number to improper fractions. Then invert and multiply.

Use These Steps

Divide $4\frac{3}{4} \div 2$

1. Change the mixed number, $4\frac{3}{4}$, and the whole number, 2, to improper fractions.

    $4\frac{3}{4} = \frac{19}{4}$ $2 = \frac{2}{1}$

2. Invert the improper fraction, $\frac{2}{1}$.

    $\frac{2}{1} \searrow \frac{1}{2}$

3. Multiply the new fractions. Change the answer to a mixed nmber.

    $\frac{19}{4} \times \frac{1}{2} = \frac{19}{8} = 2\frac{3}{8}$

Divide. Cancel and reduce if possible.

1. $6\frac{5}{8} \div 3 =$

    $\frac{53}{8} \div \frac{3}{1} = \frac{53}{8} \times \frac{1}{3} = \frac{53}{24} = 2\frac{5}{24}$

2. $5\frac{2}{3} \div 2 =$

3. $8\frac{1}{5} \div 5 =$

4. $7\frac{3}{7} \div 6 =$

5. $3\frac{1}{2} \div 4 =$

6. $2\frac{5}{6} \div 7 =$

7. $1\frac{2}{9} \div 2 =$

8. $9\frac{7}{12} \div 10 =$

9. $10\frac{9}{11} \div 5 =$

10. $7\frac{13}{16} \div 6 =$

11. Greg has to lose weight. If he loses 2 pounds each week, how many weeks will it take him to lose $10\frac{1}{2}$ pounds?

12. Carla needs to gain weight. If she gains 3 pounds each week, how many weeks will it take her to gain $12\frac{3}{4}$ pounds?

Answer_____

Answer_____

Dividing Mixed Numbers by Mixed Numbers

To divide a mixed number by a mixed number, change both mixed numbers to improper fractions. Invert and multiply.

Use These Steps

Divide $5\frac{1}{3} \div 2\frac{1}{2}$

1. Change the mixed numbers to improper fractions.

    $5\frac{1}{3} = \frac{16}{3}$ $2\frac{1}{2} = \frac{5}{2}$

2. Invert the number to the right of the division sign.

    $\frac{5}{2} \diagup \frac{2}{5}$

3. Multiply the new fractions. Change the answer to a mixed number.

    $\frac{16}{3} \div \frac{5}{2} = \frac{16}{3} \times \frac{2}{5} = \frac{32}{15} = 2\frac{2}{15}$

Divide. Cancel and reduce if possible.

1. $6\frac{1}{2} \div 1\frac{3}{4} =$
$\frac{13}{2} \div \frac{7}{4} = \frac{13}{\underset{1}{\cancel{2}}} \times \frac{\overset{2}{\cancel{4}}}{7} = \frac{26}{7} = 3\frac{5}{7}$

2. $7\frac{2}{3} \div 5\frac{1}{6} =$

3. $8\frac{3}{5} \div 3\frac{3}{10} =$

4. $10\frac{6}{7} \div 2\frac{8}{14} =$

5. $2\frac{5}{6} \div 3\frac{1}{10} =$

6. $4\frac{2}{7} \div 5\frac{1}{14} =$

7. $3\frac{5}{9} \div 4\frac{2}{27} =$

8. $12\frac{2}{5} \div 2\frac{1}{10} =$

9. $5\frac{1}{3} \div 5\frac{1}{3} =$

10. $20\frac{7}{10} \div 10\frac{1}{2} =$

11. Chris needs $3\frac{3}{4}$ yards of material to make curtains for one window in his living room. If he has $18\frac{3}{4}$ yards of material, how many windows can he make curtains for?

12. Isabel needs to run $7\frac{1}{2}$ miles every other day to get ready for a fun run. If every lap around the track is $2\frac{1}{2}$ miles, how many laps does she need to make to run $7\frac{1}{2}$ miles?

Answer _____

Answer _____

Problem Solving: Using a Sketch

Sometimes using a sketch or drawing helps you work with fractions. Hector is a carpenter. He needs to make drawings to help him remember measurements. This helps him cut the correct size pieces of wood.

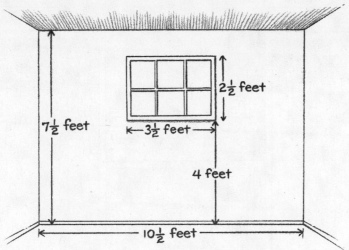

Example Hector is measuring his living room wall so that he can make bookcases to fit along that wall. He made this sketch and wrote in the measurements.

Hector wants 3 bookcases of equal size along the wall. How wide can each bookcase be?

▶ **Step 1.** Look at the drawing to find out how wide the wall is.

$$10\frac{1}{2} \text{ feet}$$

▶ **Step 2.** Divide by 3.

$$10\frac{1}{2} \div 3 = \frac{\overset{7}{\cancel{21}}}{2} \times \frac{1}{\underset{1}{\cancel{3}}} = \frac{7}{2} = 3\frac{1}{2} \text{ feet}$$

Each bookcase can be no more than $3\frac{1}{2}$ feet wide.

Solve using the sketch above.

1. Hector wants one bookcase under the window. The shelves in this bookcase are about $11\frac{1}{2}$ inches apart. How many whole shelves will Hector be able to put in this bookcase?
(Hint: 4 feet = 48 inches)

2. Hector wants one bookcase on each side of the window. Each bookcase is $6\frac{2}{3}$ feet tall. If there are 8 shelves in each bookcase, how far apart will the shelves be?

Answer_____   Answer_____

Use Hector's drawing to answer the questions.

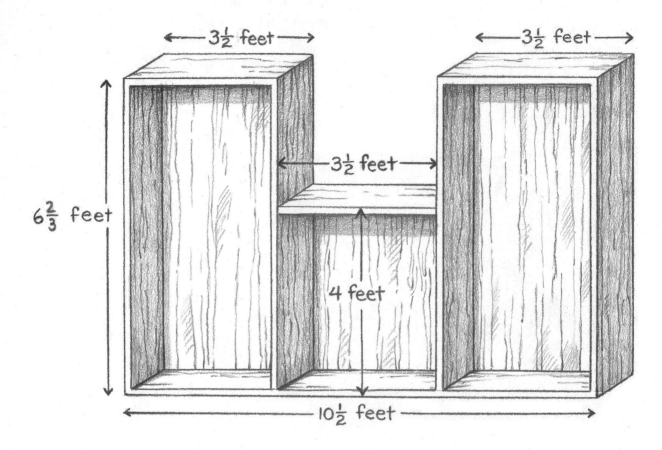

3. If Hector buys one 12-foot long board, how many whole bookcase tops can he cut from this piece?

   Answer_____

4. He buys one 14-foot long board to make the long sides of the bookcases. How many complete sides $6\frac{2}{3}$ feet long can he cut from this piece of board?

   Answer_____

5. To fit inside the two tall bookcases, the shelves must be $3\frac{5}{12}$ feet long. How many whole shelves can be cut from a 24-foot long board?

   Answer_____

6. Hector may decide to make glass doors for the middle bookcase. How wide would one of the doors be?
   (Hint: The width of the two doors together will equal $3\frac{1}{2}$ feet.)

   Answer_____

Unit 5 Review

Divide. Cancel and reduce if possible.

1. $\frac{2}{3} \div \frac{1}{2} =$
2. $\frac{3}{10} \div \frac{3}{5} =$
3. $\frac{2}{7} \div \frac{4}{9} =$
4. $\frac{5}{6} \div \frac{3}{4} =$

5. $\frac{9}{11} \div \frac{2}{3} =$
6. $\frac{5}{8} \div \frac{1}{6} =$
7. $\frac{4}{5} \div \frac{1}{5} =$
8. $\frac{2}{9} \div \frac{4}{9} =$

9. $4 \div \frac{1}{2} =$
10. $9 \div \frac{1}{3} =$
11. $7 \div \frac{3}{4} =$
12. $5 \div \frac{2}{5} =$

13. $8 \div \frac{3}{8} =$
14. $3 \div \frac{1}{7} =$
15. $6 \div \frac{5}{12} =$
16. $10 \div \frac{5}{6} =$

17. $5\frac{1}{2} \div \frac{1}{2} =$
18. $4\frac{3}{8} \div \frac{1}{4} =$
19. $6\frac{5}{9} \div \frac{2}{3} =$
20. $2\frac{2}{5} \div \frac{3}{7} =$

21. $9\frac{5}{6} \div \frac{1}{3} =$
22. $8\frac{3}{4} \div \frac{1}{2} =$
23. $10\frac{7}{10} \div \frac{1}{10} =$
24. $7\frac{2}{3} \div \frac{5}{6} =$

25. $\frac{1}{9} \div 3 =$
26. $\frac{3}{4} \div 2 =$
27. $\frac{5}{8} \div 5 =$
28. $\frac{2}{3} \div 6 =$

29. $\frac{9}{10} \div 10 =$
30. $\frac{5}{6} \div 4 =$
31. $\frac{1}{12} \div 2 =$
32. $\frac{1}{2} \div 7 =$

33. $\frac{22}{25} \div 11 =$
34. $\frac{2}{3} \div 9 =$
35. $\frac{16}{19} \div 4 =$
36. $\frac{3}{4} \div 8 =$

Divide. Cancel and reduce if possible.

37. $5\frac{1}{2} \div 4 =$

38. $1\frac{2}{5} \div 7 =$

39. $8\frac{5}{9} \div 3 =$

40. $3\frac{2}{3} \div 5 =$

41. $7\frac{3}{8} \div 2 =$

42. $6\frac{3}{10} \div 6 =$

43. $9\frac{3}{4} \div 10 =$

44. $4\frac{1}{6} \div 8 =$

45. $2\frac{7}{12} \div 5\frac{1}{8} =$

46. $10\frac{7}{9} \div 4\frac{1}{3} =$

47. $7\frac{4}{5} \div 5\frac{2}{7} =$

48. $3\frac{1}{2} \div 1\frac{1}{2} =$

49. $8\frac{5}{16} \div 2\frac{1}{16} =$

50. $15\frac{1}{6} \div 2\frac{1}{2} =$

51. $11\frac{3}{4} \div 1\frac{1}{3} =$

52. $9\frac{11}{14} \div 2\frac{1}{7} =$

53. $5\frac{1}{2} \div 3\frac{1}{4} =$

Below is a list of the problems in this review and the pages on which the skills are taught. If you missed any problems, turn to the pages listed and practice the skills. Then correct the problems you missed in the Unit Review.

Problems	Pages	Problems	Pages
1-8	129-130	25-36	137-138
9-16	131-132	37-44	139
17-24	133-134	45-53	140

Unit 6 PUTTING YOUR SKILLS TO WORK

You have studied four operations with fractions—addition, subtraction, multiplication, and division. You have applied these skills to real-life problems, and you have learned techniques for solving word problems.

In this unit, you will study how to choose the correct operation needed to solve problems. You will also learn how to solve problems that require you to use more than one operation to find the answers.

Getting Ready

You should be familiar with the skills on this page and the next before you begin this unit. To check your answers, turn to page 203.

 To add fractions and mixed numbers, be sure the denominators are the same. Add the fractions. Add the whole numbers.

Add. Reduce if possible.

1. $\frac{3}{4} = \frac{9}{12}$
    $+\frac{2}{3} = \frac{8}{12}$
    $\frac{17}{12} = 1\frac{5}{12}$

2. $1\frac{5}{8}$
    $+\ \frac{4}{8}$

3. $6\frac{2}{9}$
    $+7\frac{17}{18}$

4. $\frac{9}{10}$
    $+\frac{7}{8}$

5. $\frac{15}{21}$
    $+4\frac{2}{7}$

6. $\frac{5}{6}$
    $+\frac{1}{6}$

7. $2\frac{3}{5}$
    $+\ \frac{3}{4}$

8. $\frac{1}{5}$
    $+3\frac{2}{15}$

9. $3\frac{5}{8}$
    $+7\frac{1}{2}$

10. $\frac{8}{9}$
    $+2\frac{1}{3}$

For review, see Unit 2.

Getting Ready

 To subtract fractions and mixed numbers, be sure the denominators are the same. You may need to borrow from the whole number.

Subtract. Reduce if possible.

11. $\frac{3}{4}$
    $-\frac{1}{4}$
    $\overline{\frac{2}{4} = \frac{1}{2}}$

12. $1\frac{2}{3}$
    $-\frac{1}{6}$

13. $5\frac{9}{10}$
    $-3\frac{3}{5}$

14. 1
    $-\frac{3}{8}$

15. $\frac{5}{7}$
    $-\frac{1}{2}$

16. $1\frac{1}{8}$
    $-\frac{3}{8}$

17. $7\frac{5}{12}$
    $-\frac{5}{6}$

18. 9
    $-3\frac{7}{10}$

19. $2\frac{7}{15}$
    $-\frac{2}{3}$

20. $\frac{3}{4}$
    $-\frac{2}{5}$

For review, see Unit 3.

 To multiply fractions, multiply the numerators and then the denominators. Cancel before you multiply if possible.

Multiply. Reduce if possible.

21. $\frac{3}{\cancel{4}_2} \times \frac{\cancel{2}^1}{5} = \frac{3}{10}$

22. $1\frac{3}{8} \times \frac{8}{9} =$

23. $3\frac{2}{9} \times 3 =$

24. $6 \times \frac{1}{2} =$

25. $\frac{5}{9} \times \frac{3}{10} =$

26. $\frac{2}{3} \times 5\frac{1}{8} =$

27. $2\frac{1}{2} \times 8\frac{1}{5} =$

28. $\frac{1}{3} \times \frac{3}{4} \times \frac{5}{6} =$

For review, see Unit 4.

 To divide fractions, change any whole numbers and mixed numbers to improper fractions. Invert and multiply.

Divide. Reduce if possible.

29. $\frac{5}{6} \div \frac{1}{6} =$
    $\frac{5}{\cancel{6}_1} \times \frac{\cancel{6}^1}{1} = \frac{5}{1} = 5$

30. $8 \div \frac{1}{5} =$

31. $1\frac{7}{11} \div \frac{2}{3} =$

32. $6\frac{3}{8} \div 13 =$

33. $\frac{9}{10} \div \frac{1}{4} =$

34. $\frac{2}{9} \div \frac{5}{12} =$

35. $10\frac{1}{2} \div 2\frac{1}{4} =$

36. $10 \div 5\frac{1}{10} =$

For review, see Unit 5.

Choose an Operation: Changing Units of Measurement

To change from one unit of measurement to another, remember that when you change a small unit to a larger unit, you divide. When you change a large unit to a smaller unit, you multiply.

12 inches	=	1 foot
3 feet	=	1 yard
1,760 yards	=	1 mile
5,280 feet	=	1 mile

Example Julian bought a workbench that is $5\frac{5}{16}$ feet long. How many inches are in $5\frac{5}{16}$ feet?

▶ **Step 1.** You are changing from a large unit (feet) to a smaller unit (inches). You will need to multiply.

▶ **Step 2.** Use the chart to find out how many inches are in one foot.

$$12 \text{ inches} = 1 \text{ foot}$$

▶ **Step 3.** Multiply.

$$5\frac{5}{16} \times 12 = \frac{85}{16} \times \frac{12}{1} = \frac{255}{4} = 63\frac{3}{4}$$

The bench is $63\frac{3}{4}$ inches long.

Solve.

1. Casey cut $1\frac{1}{2}$ feet from a piece of chain. How many inches are in $1\frac{1}{2}$ feet?

   Answer_____

2. Susan is putting a fence around her garden. The fence will be $\frac{1}{4}$ mile long. How many yards are in $\frac{1}{4}$ mile?

   Answer_____

3. Patty and Lila jog every morning for $2\frac{1}{4}$ miles. Circle the expression you would use to find how many feet they jog.

   a. $2\frac{1}{4} + 5,280$
   b. $2\frac{1}{4} \times 1,760$
   c. $2\frac{1}{4} \times 5,280$
   d. $2\frac{1}{4} \div 1,760$

   Solve for the answer.

4. Circle the expression you would use to find how many yards they jog.

   a. $2\frac{1}{4} \times 5,280$
   b. $2\frac{1}{4} \times 1,760$
   c. $2\frac{1}{4} \div 5,280$
   d. $2\frac{1}{4} - 1,760$

   Solve for the answer.

   Answer_____   Answer_____

Choose an Operation: Being a Consumer

Comparing unit costs is one way to find out which of two brands is less expensive to buy. To find the unit cost, write a fraction for each brand.

Example Tasty green beans are 2 cans for 43 cents. Flavorite green beans are 3 cans for 65 cents. Which is less expensive?

▶ **Step 1.** Find the unit cost for each brand by writing a fraction for each. Write the cost as the numerator and the unit as the denominator. Change each to a mixed number. Find a common denominator for the fractions.

Tasty: $\frac{43}{2} = 21\frac{1}{2} = 21\frac{3}{6}$ Flavorite: $\frac{65}{3} = 21\frac{2}{3} = 21\frac{4}{6}$

▶ **Step 2.** Compare the costs.

$21\frac{4}{6} > 21\frac{3}{6}$, so $21\frac{2}{3} > 21\frac{1}{2}$
Tasty is less expensive.

Solve.

1. Crispy Apple Bites cost 79 cents for the 10-ounce package. Crispy Cherry Bites cost 89 cents for the 11-ounce package. Which costs more?

   Answer_____

2. Yellow apples are 4 for 99 cents. Red apples are 3 for 69 cents. Which cost less?

   Answer_____

3. Beefy hot dogs are 7 packages for 6 dollars. Meaty hot dogs are 6 packages for 5 dollars. Which cost less?

   Answer_____

4. Softy tissues are on sale 3 boxes for 2 dollars. Squeezy tissues are on sale 8 boxes for 6 dollars. Which cost more?

   Answer_____

Choose an Operation: Using a Line Graph

Carl has worked for A-Mart for five years and is eligible to buy stock in the company. He decided to watch how the stock was doing on the stock market for a week before he made up his mind whether to buy any. He learned that stocks are sold in shares and prices are given in units called points.

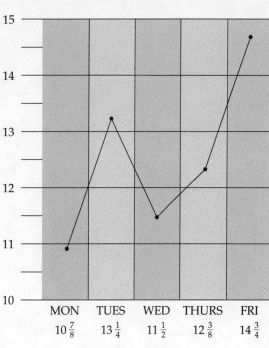

Example How much did the price of a share of stock go up from Monday to Tuesday?

▶ **Step 1.** Find the price of the stock on Monday, $10\frac{7}{8}$. Find the price of the stock on Tuesday, $13\frac{1}{4}$.

▶ **Step 2.** Subtract to find the difference.

$$13\frac{1}{4} = 13\frac{2}{8} = 12\frac{10}{8}$$
$$-10\frac{7}{8} = 10\frac{7}{8} = 10\frac{7}{8}$$
$$\phantom{-10\frac{7}{8} = 10\frac{7}{8} = 10\,} 2\frac{3}{8}$$

The price of one share of stock went up $2\frac{3}{8}$ points from Monday to Tuesday.

Solve.

1. How many points did a share of stock go down from Tuesday to Wednesday?

    Answer_____

2. How much higher was the stock on Thursday than it was on Monday?

    Answer_____

3. Circle the expression you would use to find the price of 20 shares of stock if you bought them on Tuesday.

    a. $20 + 13\frac{1}{4}$ Solve for the answer.
    b. $20 - 13\frac{1}{4}$
    c. $20 \times 13\frac{1}{4}$
    d. $20 \div 13\frac{1}{4}$

4. Circle the expression you would use to find the price of one share of stock if the price went up $1\frac{1}{8}$ points from the price it was on Friday.

    a. $14\frac{3}{4} + 1\frac{1}{8}$ Solve for the answer.
    b. $14\frac{3}{4} - 1\frac{1}{8}$
    c. $14\frac{3}{4} \times 1\frac{1}{8}$
    d. $14\frac{3}{4} \div 1\frac{1}{8}$

Answer_____

Answer_____

Multi-Step Problems: Using Measurement

You may need to use two or more operations to solve a problem. When working with measurement, you may need to add, subtract, multiply, or divide. You may also need to change from one unit to another unit.

1 cup	=	8 ounces
1 pint	=	2 cups
1 quart	=	2 pints
1 gallon	=	4 quarts

Example Lucy is changing the oil in her car. She has $\frac{3}{4}$ gallon of oil. She needs 6 quarts in all. How many more quarts does she need?

▶ **Step 1.** Change $\frac{3}{4}$ gallon to quarts. To change from a large unit to a smaller unit, multiply.

$$\frac{3}{4} \times 4 = \frac{3}{4} \times \frac{4}{1} = \frac{3}{1} = 3$$

Lucy has 3 quarts.

▶ **Step 2.** To find out how many more quarts she needs, subtract.

$$6 - 3 = 3$$

Lucy needs 3 more quarts.

Solve.

1. Pete is painting a chest of drawers. He needs $\frac{1}{2}$ gallon of paint. He has 1 quart. How many more quarts does he need?

   Answer_____

2. Each spring Tony uses $3\frac{1}{2}$ gallons of fertilizer for his yard. If he makes 12 quarts of fertilizer from concentrate, how many more quarts will he need?

   Answer_____

3. Dawn is making cheese puffs. She needs 3 cups of milk. She has $\frac{1}{2}$ pint. How many more cups does she need?

   Answer_____

4. Norm is planning to buy some wood sealer to seal his deck. He has $\frac{1}{2}$ gallon in one can and a full quart can. How many quarts of sealer does he have all together?

   Answer_____

Solve.

5. Danielle filled 12 pint containers and 20 half-pint containers with green beans for her freezer. How many pints did she freeze all together?

   Answer_____

6. Jennifer is using a recipe that calls for 24 ounces of sour cream. She has $1\frac{1}{2}$ cups of sour cream in her refrigerator. How many more cups does she need to buy?

   Answer_____

7. David is making 3 pecan pies. He needs $1\frac{1}{2}$ cups of pecans for each pie. How many ounces of pecans does he need for all 3 pies?

   Answer_____

8. Randi bought two 8-ounce packages of cream cheese. She used $1\frac{3}{4}$ cups to make sandwiches. How much cream cheese does she have left?

   Answer_____

9. Delores makes 32 ounces of fresh orange juice every Saturday morning for her family. If each serving of juice is $\frac{1}{2}$ cup, how many servings will Delores get from 32 ounces?

   Answer_____

10. Frank is repainting his house. He needs $1\frac{2}{3}$ quarts of paint for each room in the house. If he buys $2\frac{1}{2}$ gallons of paint, how many rooms can he expect to paint?

    Answer_____

11. Gwen buys coffee beans in 22-ounce packages. She uses $\frac{1}{4}$ cup of beans to make each pot of coffee. How many pots of coffee can she make from one 22-ounce package?

    Answer_____

12. Liz works as a dietary aide in a nursing home. She gives patients fresh fruit salad in $\frac{1}{2}$-cup servings. How many servings will she get from $1\frac{1}{2}$ quarts of fruit salad?

    Answer_____

Multi-Step Problems: Being a Consumer

If you buy 3 packs of gum for a dollar, you can write this as a fraction, $\frac{3}{1}$. If you buy 3 pounds of fish for $5, the fraction is $\frac{3}{5}$. These fractions are called ratios.

Ratios are useful for solving some kinds of problems.

Example   Bert's Roadside Stand sells walnuts in 10-pound bags for $25. How many pounds will you get for $5?

Step 1.   Write a ratio (fraction) to show 10 pounds for $25.

$$\frac{10 \text{ pounds}}{25 \text{ dollars}}$$

Step 2.   Write a second ratio (fraction) to show the unknown number of pounds for $5.

$$\frac{? \text{ pounds}}{5 \text{ dollars}}$$

Step 3.   Write the two ratios as equal fractions. This is called a proportion.

$$\frac{10}{25} = \frac{?}{5}$$

Step 4.   Find the numerator of the second fraction. Since $25 \div 5 = 5$, divide the numerator by 5.

$$\frac{10}{25} = \frac{10 \div 5}{25 \div 5} = \frac{2}{5}$$

You can get 2 pounds of walnuts for $5.

Solve.

1. Bert's Roadside Stand sells 20 pounds of potatoes for $4. How many pounds can you get for $2?

   Answer _____

2. If onions cost $2 for 4 pounds, how many pounds can you get for a dollar?

   Answer _____

3. Bert's Roadside Stand sells 2 jars of strawberry jam for $3. How much will it cost to buy 4 jars?
   (Hint: Find the denominator.)

   Answer _____

4. Apple cider costs $5 for 2 half-gallons. How much do 6 half-gallons cost?

   Answer _____

5. If oranges are 5 pounds for $4, how much do 15 pounds cost?

   Answer _____

6. Homemade cherry pies are 3 for $10. How many pies will you get for $20?

   Answer _____

7. Asparagus is $2 a pound. How many pounds can you get for $8?

   Answer _____

8. Homemade pickles cost $5 for 3 jars. How much will 6 jars cost?

   Answer _____

9. Bert's sells 6 dozen fresh eggs for $4. How much would 3 dozen eggs cost?

   Answer _____

10. Honey costs $4 for 3 pints. How many pints will you get for $12?

    Answer _____

11. Carrots are 3 bunches for a dollar. How much will 6 bunches of carrots cost?

    Answer _____

12. Fresh corn costs a dollar for 6 ears. How much would 3 dozen ears cost?
    (Hint: 1 dozen = 12)

    Answer _____

Multi-Step Problems: Using a Time Record

Augusta offers child care in her own home on a drop-in basis. She must keep a record of how much time each child spends in her care. She charges parents based on these records.

	Child	Mon.	Tues.	Wed.	Thurs.	Fri.
1	Teddy	7:30–11:45	8–11:30	Sick	Sick	
2	Peter H.	1–3:45	12:30–6:30	1:15–5:30	2–6:30	—
3	Susie	5:30–10:45	—	6–11:30	—	5:45–10:45
4	Peter W.					
5	Hannah					
6						
7						
8						
9						

Example On Monday, Teddy's mother dropped him off at 7:30 and picked him up at 11:45. How many hours did Teddy spend at Augusta's on Monday?

▶ **Step 1.** To find the number of hours Teddy spent at Augusta's on Monday, subtract the time he arrived from the time he left.

$$\begin{array}{r} 11 \text{ hours } \ 45 \text{ minutes} \\ - \ 7 \text{ hours } \ 30 \text{ minutes} \\ \hline 4 \text{ hours } \ 15 \text{ minutes} \end{array}$$

▶ **Step 2.** Write the time as a mixed number with 60 minutes as the denominator of the fraction. Reduce.

$$4 \text{ hours } 15 \text{ minutes} = 4\frac{15}{60} = 4\frac{1}{4}$$

Teddy spent $4\frac{1}{4}$ hours at Augusta's on Monday.

Solve. Put your answers in the correct space on the time record.

1. How many hours did Teddy spend at Augusta's on Tuesday?

   Answer_____

2. Teddy stayed at Augusta's from 8:00 to 12:30 on Friday morning. Write these times in the chart. How many hours did Teddy stay on Friday morning?

   Answer_____

3. Peter W. stayed at Augusta's only on Thursday from 1:30 to 3:45. Fill in the weekly record for Peter W. on the chart. How much time did Peter W. spend at Augusta's on Thursday?

   Answer_____

4. Susie's mother works Monday, Wednesday, and Friday mornings. Susie stays at Augusta's only on these days. How many hours did Susie stay on Wednesday?

   Answer_____

5. Hannah's dad dropped her off at Augusta's on Monday at 3:15 and her mom picked her up at 6:45. Write these times in the chart. How many hours did Hannah spend at Augusta's on Monday?

   Answer_____

6. Hannah was dropped off at Augusta's again on Friday at 3:30 and she was picked up at 6:30. Write these times in the chart. How many hours did Hannah stay on Friday?

   Answer_____

7. Augusta charges $4 for each hour of child care. Peter H. stayed $4\frac{1}{4}$ hours on Wednesday. How much did Augusta charge Peter's parents for Wednesday?

   Answer_____

8. How many hours did Peter H. stay at Augusta's on Monday? How much did Augusta charge his parents for Monday?

   Answer_____

Fractions Skills Inventory

Reduce to lowest terms. If the fraction is in lowest terms, write LT.

1. $\dfrac{3}{6} =$
2. $\dfrac{3}{12} =$
3. $\dfrac{3}{20} =$
4. $\dfrac{30}{40} =$
5. $\dfrac{24}{36} =$

Raise to higher terms with the given denominator.

6. $\dfrac{1}{3} = \dfrac{}{12}$
7. $\dfrac{2}{7} = \dfrac{}{21}$
8. $\dfrac{1}{2} = \dfrac{}{10}$
9. $\dfrac{4}{5} = \dfrac{}{20}$
10. $\dfrac{5}{8} = \dfrac{}{32}$

Change to a whole or mixed number.

11. $\dfrac{5}{4} =$
12. $\dfrac{10}{5} =$
13. $\dfrac{15}{7} =$
14. $\dfrac{4}{1} =$
15. $\dfrac{21}{10} =$

Change to an improper fraction.

16. $2\dfrac{2}{5} =$
17. $3\dfrac{1}{3} =$
18. $5\dfrac{3}{4} =$
19. $1\dfrac{2}{9} =$
20. $8\dfrac{1}{12} =$

Add. Reduce if possible.

21. $\dfrac{1}{4} + \dfrac{1}{4} =$
22. $\dfrac{3}{5} + \dfrac{2}{5} =$
23. $\dfrac{5}{6} + \dfrac{4}{6} =$
24. $\dfrac{3}{8} + \dfrac{1}{8} + \dfrac{5}{8} =$

25. $3\dfrac{5}{7} \\ +\ 1\dfrac{1}{7}$
26. $4\dfrac{1}{10} \\ +\ 6\dfrac{3}{10}$
27. $9\dfrac{2}{3} \\ +\ 3\dfrac{2}{3}$
28. $\dfrac{2}{9} \\ +\ 5\dfrac{7}{9}$
29. $1 \\ +\ 6\dfrac{3}{4}$

30. $\dfrac{1}{3} \\ +\ \dfrac{1}{6}$
31. $\dfrac{1}{4} \\ +\ \dfrac{3}{5}$
32. $\dfrac{2}{7} \\ +\ \dfrac{9}{21}$
33. $\dfrac{5}{8} \\ +\ \dfrac{3}{10}$
34. $\dfrac{5}{9} \\ +\ \dfrac{3}{4}$

Add. Reduce if possible.

35. $5\frac{5}{7}$
$+2\frac{1}{3}$

36. 8
$+2\frac{3}{8}$

37. $10\frac{9}{14}$
$+3\frac{1}{2}$

38. $12\frac{4}{5}$
$+3\frac{3}{20}$

39. $\frac{5}{6}$
$+5\frac{1}{8}$

Subtract. Reduce if possible.

40. $\frac{7}{8} - \frac{3}{8} =$

41. $\frac{5}{9} - \frac{1}{9} =$

42. $\frac{3}{7} - \frac{3}{7} =$

43. $\frac{7}{10} - \frac{1}{10} =$

44. $8\frac{5}{6}$
$-2\frac{1}{6}$

45. $6\frac{4}{5}$
$-\frac{2}{5}$

46. $7\frac{3}{4}$
$-3\frac{1}{4}$

47. $2\frac{13}{15}$
$-\frac{7}{15}$

48. $9\frac{10}{21}$
-3

49. 1
$-\frac{2}{7}$

50. 5
$-\frac{2}{9}$

51. $7\frac{3}{10}$
$-\frac{9}{10}$

52. $6\frac{1}{8}$
$-4\frac{3}{8}$

53. $1\frac{5}{12}$
$-\frac{7}{12}$

54. $\frac{2}{3}$
$-\frac{1}{6}$

55. $\frac{5}{12}$
$-\frac{1}{4}$

56. $\frac{7}{8}$
$-\frac{3}{10}$

57. $\frac{3}{4}$
$-\frac{1}{7}$

58. $\frac{4}{5}$
$-\frac{2}{3}$

59. $5\frac{1}{3}$
$-2\frac{2}{9}$

60. $9\frac{2}{7}$
$-\frac{2}{5}$

61. $8\frac{3}{16}$
$-4\frac{5}{8}$

62. $1\frac{1}{6}$
$-\frac{1}{4}$

Multiply. Reduce if possible.

63. $\frac{2}{5} \times \frac{1}{2} =$

64. $\frac{5}{6} \times \frac{3}{10} =$

65. $\frac{3}{8} \times \frac{4}{7} =$

66. $\frac{2}{9} \times \frac{6}{8} \times \frac{4}{5} =$

67. $8 \times \frac{1}{4} =$

68. $\frac{4}{9} \times 6 =$

69. $3\frac{1}{5} \times 10 =$

70. $5 \times 1\frac{2}{7} =$

71. $4\frac{9}{10} \times \frac{5}{7} =$

72. $\frac{3}{5} \times 3\frac{3}{4} =$

73. $2\frac{1}{3} \times 5\frac{4}{7} =$

74. $1\frac{1}{5} \times 6\frac{1}{2} =$

Divide. Reduce if possible.

75. $\frac{3}{4} \div \frac{1}{4} =$

76. $\frac{1}{2} \div \frac{3}{8} =$

77. $\frac{5}{9} \div \frac{9}{10} =$

78. $\frac{2}{3} \div \frac{2}{7} =$

79. $8 \div \frac{1}{2} =$

80. $7 \div \frac{1}{6} =$

81. $4\frac{3}{7} \div \frac{5}{14} =$

82. $3\frac{2}{5} \div \frac{2}{5} =$

83. $\frac{2}{3} \div 2 =$

84. $\frac{1}{5} \div 5 =$

85. $1\frac{8}{9} \div 3 =$

86. $4\frac{1}{6} \div 2 =$

87. $10\frac{1}{2} \div 5 =$

88. $4\frac{2}{3} \div 2\frac{1}{3} =$

89. $1\frac{7}{8} \div 4\frac{3}{8} =$

90. $25\frac{1}{2} \div 3\frac{1}{4} =$

Below is a list of the problems in this Skills Inventory and the pages on which the skills are taught. If you missed any problems, turn to the pages listed and practice the skills. Then correct the problems you missed in the Skills Inventory.

Problem	Practice Page
Unit 1	
1-5	15-16
6-10	19-20
11-15	33-34
16-20	35-36
Unit 2	
21-24	41-42, 44-46
25-29	49-51
30-34	56-59, 63-65
35-39	67-68

Problem	Practice Page
Unit 3	
40-43	75-77
44-48	78, 80-81
49-53	86-90
54-58	93-96
59-62	100-103

Problem	Practice Page
Unit 4	
63-66	109-110, 112-115
67-70	117-119
71-74	122-123
Unit 5	
75-78	129-130
79-82	131-134
83-87	137-139
88-90	140

Glossary

addition (page 39) - Putting numbers together to find a total. The symbol + is used in addition.

$$\begin{array}{r}\frac{1}{5}\\+\frac{2}{5}\\\hline\frac{3}{5}\end{array}$$

borrowing (page 87) - Taking an amount from the whole number and adding it to the fraction part of a mixed number.

$$4\frac{1}{3} = 3\frac{1}{3} + \frac{3}{3} = 3\frac{4}{3}$$
$$-1\frac{2}{3} = 1\frac{2}{3}$$
$$ 2\frac{2}{3}$$

cancellation (page 113) - Dividing a numerator and a denominator by the same number when you multiply or divide fractions.

$$\frac{\cancel{4}^{2}}{5} \times \frac{1}{\cancel{2}_{1}} = \frac{2}{5}$$

chart (page 43) - Information arranged in rows and columns.

ITEM	NUMBER SOLD	FRACTION OF TOTAL
Rulers	8	$\frac{1}{6}$
Scissors	2	$\frac{1}{24}$
Pencils	16	$\frac{1}{3}$
Staplers	4	$\frac{1}{12}$
Pens	12	$\frac{1}{4}$
Notepads	6	$\frac{1}{8}$
TOTAL ITEMS SOLD	48	

circle graph (page 17) - A circle cut into sections to show the parts that make a total. Each part is a fraction of the total.

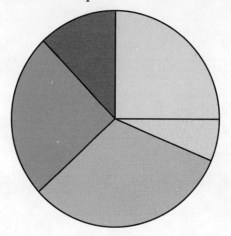

common denominator (page 25) - A number that all denominators you're working with divide into evenly.

$$\frac{1}{2} = \frac{3}{6}$$
$$\frac{1}{3} = \frac{2}{6}$$

comparing (page 23) - Deciding if a fraction is equal to, greater than, or less than another fraction.

denominator (page 11) - The bottom number in a fraction. The number of equal parts in the whole.

$$\frac{2}{3}$$

difference (page 73) - The answer to a subtraction problem.

$$\begin{array}{r}\frac{3}{8}\\-\frac{2}{8}\\\hline\frac{1}{8}\end{array}$$

division (page 127) - Splitting an amount into equal groups. The symbol ÷ is used in division.

$$\frac{1}{4} \div \frac{2}{3} = \frac{1}{4} \times \frac{3}{2} = \frac{3}{8}$$

equal (page 11) - The same in value. The symbol = means *equal*.

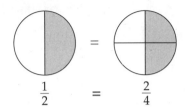

fraction (page 11) - Part of a whole or a group.

 = $\frac{3}{4}$

fraction bar (page 11) - The line that separates the numerator and denominator of a fraction.
$\frac{2}{3}$

greater than (page 23) - More than. The symbol > means *greater than*.
$\frac{2}{7} > \frac{1}{7}$ means $\frac{2}{7}$ is greater than $\frac{1}{7}$

horizontal (page 42) - Side-to-side.

---

higher terms (page 19) - A fraction is in higher terms when you multiply the numerator and denominator by the same number.
$\frac{1}{2} = \frac{1 \times 2}{2 \times 2} = \frac{2}{4}$

improper fraction (page 31) - A fraction with the numerator equal to or larger than the denominator.
$\frac{3}{3} \quad \frac{4}{3}$

invert (page 129) - Turn upside down.
$\frac{1}{2} \bowtie \frac{2}{1}$

less than (page 23) - Smaller than. The symbol < means *less than*.
$\frac{6}{9} < \frac{8}{9}$ means $\frac{6}{9}$ is less than $\frac{8}{9}$

line graph (page 149) - A graph with lines connecting points that show different amounts.

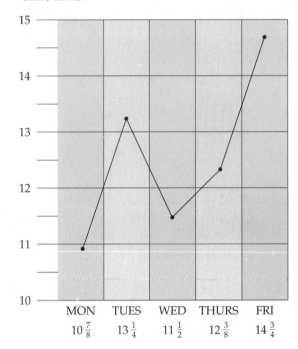

lowest common denominator (LCD) (page 65) - The smallest number that the denominators of fractions being added or subtracted divide into evenly.
$\frac{1}{4} = \frac{3}{12}$
$\frac{1}{3} = \frac{2}{12}$

lowest terms (page 15) - A fraction is in lowest terms when 1 is the only number that divides evenly into both the numerator and denominator.
$\frac{4}{8} = \frac{4 \div 4}{8 \div 4} = \frac{1}{2}$

mixed number (page 26) - A number with a whole number part and a fraction part.
$3\frac{1}{10}$

multiple (page 65) - Number found when you multiply a whole number by 0, 1, 2, 3, and so on.
$3 \times 1 = 3 \quad 3 \times 2 = 6$
$3 \times 3 = 9 \quad 3 \times 4 = 12$

multiplication (page 107) - Combining equal numbers two or more times to get a total. The symbol × is used in multiplication.
$\frac{2}{3} \times \frac{1}{3} = \frac{2}{9}$

number line (page 23) - A line with equally spaced points that are labeled with fractions, whole numbers, or mixed numbers.

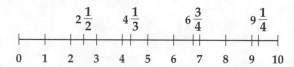

numerator (page 11) - The top number in a fraction. The number of parts being considered.
$\frac{5}{9}$

of (page 109) - *Of* means to multiply or find part of something.
$\frac{1}{2}$ of $\frac{9}{10} = \frac{1}{2} \times \frac{9}{10}$

operation (page 145) - The process you use to solve a math problem. The basic operations are addition, subtraction, multiplication, and division.

proper fraction (page 31) - A fraction with a numerator that is smaller than the denominator.
$\frac{2}{3}$

proportion (page 152) - Two equal ratios or fractions.
$\frac{2}{3} = \frac{4}{9}$

ratio (page 152) - A fraction showing the relationship of two numbers.
$\frac{4}{6} \begin{array}{l} \text{pounds} \\ \text{dollars} \end{array}$

reducing (page 15) - Dividing both the numerator and denominator of a fraction by the same number.
$\frac{3}{9} = \frac{3 \div 3}{9 \div 3} = \frac{1}{3}$

remainder (page 27) - The numerator of a fraction when the amount left over in a division problem is shown as a fraction.
$\begin{array}{r} 6 \text{ R } 1 = 6\frac{1}{6} \\ 6\overline{)37} \\ -36 \\ \hline 1 \end{array}$

subtraction (page 73) - Taking away a certain amount from another amount to find a difference. The symbol − is used in subtraction.
$\begin{array}{r} \frac{5}{9} \\ -\frac{2}{9} \\ \hline \frac{3}{9} \end{array}$

sum (page 35) - The answer to an addition problem.
$\begin{array}{r} \frac{2}{7} \\ +\frac{3}{7} \\ \hline \frac{5}{7} \end{array}$

table (page 18) - Information arranged in rows and columns.

ITEM	NUMBER SOLD	FRACTION OF TOTAL
Rulers	8	$\frac{1}{6}$
Scissors	2	$\frac{1}{24}$
Pencils	16	$\frac{1}{3}$
Staplers	4	$\frac{1}{12}$
Pens	12	$\frac{1}{4}$
Notepads	6	$\frac{1}{8}$
TOTAL ITEMS SOLD	48	

vertical (page 42) - Up-and-down.

|

whole number (page 26) - A number that shows a whole amount.

$6 \frac{5}{8}$

Answers & Explanations

The answer to the problem that was worked out for you in the lesson is written here in color. The next answer has an explanation written beneath it. The answers to the rest of the problems in the lesson follow in order.

Skills Inventory

Page 6

1. $\frac{1}{2}$
2. $\frac{3}{4}$
3. LT
4. $\frac{1}{3}$
5. $\frac{2}{3}$
6. $\frac{3}{9}$
7. $\frac{4}{10}$
8. $\frac{9}{21}$
9. $\frac{21}{30}$
10. $\frac{16}{36}$
11. $1\frac{1}{3}$
12. 4
13. $2\frac{2}{5}$
14. 3
15. $1\frac{4}{5}$
16. $\frac{23}{4}$
17. $\frac{5}{2}$
18. $\frac{14}{3}$
19. $\frac{39}{5}$
20. $\frac{97}{10}$
21. $\frac{3}{5}$
22. $\frac{3}{4}$
23. $1\frac{3}{5}$
24. 1
25. $7\frac{5}{9}$
26. $9\frac{1}{2}$
27. $8\frac{2}{3}$
28. 10
29. $15\frac{11}{15}$
30. $1\frac{1}{4}$
31. $1\frac{5}{21}$
32. $\frac{20}{27}$
33. $\frac{13}{24}$
34. $1\frac{11}{20}$

Page 7

35. $3\frac{25}{42}$
36. $13\frac{1}{2}$
37. $16\frac{2}{15}$
38. 11
39. $7\frac{11}{15}$
40. $\frac{2}{5}$
41. $\frac{2}{3}$
42. 0
43. $\frac{1}{2}$
44. $7\frac{2}{11}$
45. $8\frac{1}{5}$
46. $1\frac{2}{3}$
47. $9\frac{2}{3}$
48. $7\frac{17}{25}$
49. $\frac{2}{5}$
50. $3\frac{1}{8}$
51. $6\frac{2}{3}$
52. $\frac{4}{7}$
53. $\frac{5}{9}$
54. $\frac{1}{4}$
55. $\frac{1}{10}$
56. $\frac{7}{24}$
57. $\frac{1}{12}$
58. $\frac{11}{42}$
59. $2\frac{1}{2}$
60. $4\frac{13}{20}$
61. $\frac{43}{60}$
62. $\frac{11}{18}$

Page 8

63. $\frac{1}{3}$
64. $\frac{1}{6}$
65. $\frac{5}{8}$
66. $\frac{1}{10}$
67. $\frac{2}{3}$
68. $1\frac{1}{2}$
69. 34
70. $25\frac{2}{3}$
71. $\frac{7}{9}$
72. $1\frac{3}{5}$
73. 6
74. $22\frac{1}{2}$
75. $2\frac{2}{3}$
76. 5
77. $1\frac{1}{9}$
78. 2
79. 27
80. 16
81. $31\frac{1}{2}$
82. 12
83. $\frac{1}{6}$
84. $\frac{2}{21}$
85. $\frac{19}{32}$
86. $3\frac{1}{10}$
87. $1\frac{4}{5}$
88. 3
89. $2\frac{11}{12}$
90. $6\frac{4}{21}$

Unit 1

Page 9

1. 70
2. 51

    19
    + 32

    51

3. 162
4. 143

5.	925	6.	395
7.	520	8.	840

Page 10
9. 27
10. 25

```
  41
- 16
-----
  25
```

11.	35	12.	68

13. **66**
14. 276

```
  46
×  6
-----
 276
```

15.	122	16.	415

17. **13 R1**
18. 19

```
    19
2)38
  - 2
  ---
   18
 - 18
  ---
    0
```

19.	5 R4	20.	7

Page 11
1. $\frac{4}{8}$
2. $\frac{2}{4}$
   2 of the 4 parts are shaded.

3.	$\frac{2}{5}$	4.	$\frac{1}{3}$
5.	$\frac{5}{9}$	6.	$\frac{7}{10}$
7.	$\frac{8}{10}$	8.	$\frac{3}{4}$

9.

10.–12. *Shading should be similar to this.*
10.

2 of the 3 parts are shaded.

11.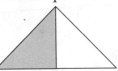

12.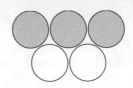

Page 12
1. one half
2. five sixths
   There are 6 circles in the whole. 5 of them are shaded.

3.	one third	4.	one fifth
5.	two thirds	6.	four ninths
7.	three sevenths	8.	seven tenths
9.	one half	10.	three eighths
11.	five eighths	12.	four fourths

Page 13
1. $\frac{3}{12}$
2. $\frac{25}{36}$
   Write the parts in a whole yard, 36 inches, as the denominator. Write the number of parts given, 25 inches, as the numerator.

3.	$\frac{1}{4}$	4.	$\frac{2}{3}$
5.	$\frac{9}{10}$	6.	$\frac{42}{60}$
7.	$\frac{17}{24}$	8.	$\frac{12}{12}$

Page 14
1. Cake C
   $\frac{3}{5}$
2. Cake A
   $\frac{3}{8}$
   Cake A is cut into 8 equal parts (eighths). 5 of the 8 parts are still in the pan. 3 of the 8 parts have been eaten. $\frac{3}{8}$ of the cake has been eaten.

3.	$\frac{3}{6}$	4.	$\frac{1}{4}$
5.	$\frac{6}{10}$	6.	$\frac{5}{8}$

Page 15
1. $\frac{2}{3}$
2. $\frac{3}{7}$
   Divide each number by 3. The fraction is reduced to $\frac{3}{7}$. Only 1 divides into 3 and 7 evenly. $\frac{3}{7}$ is in lowest terms.
   $\frac{9}{21} = \frac{9 \div 3}{21 \div 3} = \frac{3}{7}$

3. $\frac{7}{10}$ 4. $\frac{3}{4}$
5. $\frac{5}{8}$ 6. $\frac{4}{5}$
7. $\frac{3}{4}$ 8. $\frac{5}{6}$
9. $\frac{3}{4}$ 10. $\frac{9}{10}$
11. $\frac{4}{5}$ 12. $\frac{2}{3}$
13. $\frac{1}{3}$ 14. $\frac{5}{6}$
15. $\frac{1}{2}$ 16. $\frac{3}{5}$
17. $\frac{1}{2}$ 18. $\frac{1}{3}$

Page 16

1. $\frac{3}{5}$
2. LT
   Only 1 divides evenly into 9 and 25.
3. $\frac{2}{3}$ 4. $\frac{1}{4}$
5. $\frac{3}{7}$ 6. $\frac{1}{10}$
7. LT   8. $\frac{30}{73}$
9. LT
10. $\frac{1}{3}$
    $\frac{8}{24} = \frac{8 \div 8}{24 \div 8} = \frac{1}{3}$
11. $\frac{2}{3}$ 12. LT
13. LT   14. $\frac{1}{7}$
15. LT   16. $\frac{3}{5}$
17. $\frac{1}{2}$
    Set up a fraction with the total gallons the gas tank will hold, 20, as the denominator. The number of gallons of gas Jack put into the tank, 10, is the numerator. Reduce the fraction to lowest terms by crossing out the zeros on the top and the bottom.
    $\frac{1\cancel{0}}{2\cancel{0}} = \frac{1}{2}$
18. $\frac{3}{4}$
    Reduce to lowest terms by dividing the numerator and the denominator by 2.
    $\frac{6}{8} = \frac{6 \div 2}{8 \div 2} = \frac{3}{4}$

Page 17

1. $\frac{1}{4}$
   Set up a fraction with the total expenses, $1,200, as the denominator. The amount spent for employees' salaries, $300, is the numerator. Cross out the same number of zeros on the end of the numerator and the denominator. Reduce to lowest terms by dividing the numerator and denominator by 3.
   $\frac{3\cancel{00}}{12\cancel{00}} = \frac{3}{12} = \frac{3 \div 3}{12 \div 3} = \frac{1}{4}$
2. $\frac{1}{2}$
   $\frac{6\cancel{00}}{12\cancel{00}} = \frac{6 \div 6}{12 \div 6} = \frac{1}{2}$

Page 18

3. $\frac{1}{12}$
   $\frac{1\cancel{00}}{12\cancel{00}} = \frac{1}{12}$
4. $\frac{7}{12}$
   Add the amount spent on car parts, $100, and the amount spent on gas and oil, $600. $100 + $600 = $700. Set up a fraction with the total, $700, as the numerator. Manuel's total expenses, $1,200, is the denominator. Reduce by crossing out the same number of zeros on the end of the numerator and the denominator.
   $\frac{7\cancel{00}}{12\cancel{00}} = \frac{7}{12}$
5. $\frac{5}{12}$
   Rent ($200) + Salaries ($300) = $500
   $\frac{5\cancel{00}}{12\cancel{00}} = \frac{5}{12}$
6. $\frac{12}{12} = 1$
   Rent $ 200
   Salaries   300
   Car parts   100
   Gas and oil   600
                $1,200
   $\frac{12\cancel{00}}{12\cancel{00}} = \frac{12}{12} = 1$
7. $\frac{1}{2}$
   $\frac{150}{300}$ mechanic's salary
                      total salary expense
   $\frac{15\cancel{0}}{30\cancel{0}} = \frac{15}{30} = \frac{1}{2}$
8. $\frac{1}{5}$
   $\frac{250}{1250}$ rent
                       total expenses
   $\frac{25\cancel{0}}{125\cancel{0}} = \frac{25}{125} = \frac{1}{5}$

9.

Expense	Amount	Fraction
rent	$200	$\frac{200}{1200} = \frac{1}{6}$
salaries	$300	$\frac{300}{1200} = \frac{1}{4}$
car parts	$100	$\frac{100}{1200} = \frac{1}{12}$
gas and oil	$600	$\frac{600}{1200} = \frac{1}{2}$

Page 19

1. $\frac{2}{3} = \frac{4}{6}$
2. $\frac{1}{2} = \frac{3}{6}$
   $\frac{1}{2} = \frac{1 \times 3}{2 \times 3} = \frac{3}{6}$
3. $\frac{1}{6} = \frac{2}{12}$
4. $\frac{1}{2} = \frac{5}{10}$
5. $\frac{7}{8} = \frac{14}{16}$
6. $\frac{3}{4} = \frac{6}{8}$
7. $\frac{8}{20}$
8. $\frac{12}{32}$
   $\frac{3}{8} = \frac{3 \times 4}{8 \times 4} = \frac{12}{32}$
9. $\frac{4}{24}$
10. $\frac{16}{20}$
11. $\frac{15}{20}$
12. $\frac{20}{35}$
    $\frac{4}{7} = \frac{4 \times 5}{7 \times 5} = \frac{20}{35}$
13. $\frac{10}{15}$
14. $\frac{5}{45}$

Page 20

1. 3
2. 4
   Both the numerator, 3, and the denominator, 4, were multiplied by 4.
3. 3
4. 6
5. 3
6. 4
7. 5
8. 5
9. $\frac{9}{21}$
10. $\frac{8}{56}$
    The denominator, 14, was multiplied by 4 to get 56. Multiply 2 by 4 to get 8.
11. $\frac{60}{100}$
12. $\frac{27}{48}$
13. $\frac{36}{90}$
14. $\frac{72}{90}$
15. $\frac{6}{54}$
16. $\frac{49}{77}$
17. $\frac{42}{84}$
18. $\frac{20}{96}$
19. $\frac{9}{63}$
20. $\frac{48}{72}$

21. 20 pages
    Write a fraction with the number of pages Kevin read, 10, as the numerator. The total number of pages in the book, 25, is the denominator. Raise this fraction, $\frac{10}{25}$, to higher terms. Use the total number of pages in the book Beverly read, 50, as the denominator.
    $\frac{10}{25} = \frac{}{50}$
    25 multiplied by 2 is 50.
    10 multiplied by 2 is 20.
    $\frac{10}{25} = \frac{20}{50}$

22. 28 pages
    Set up a fraction with the number of pages Carmen typed, 14, as the numerator. The total number of pages in the term paper, 30, is the denominator. This fraction, $\frac{14}{30}$, equals the fraction set up with the total number of pages in the paper Scott typed, 60, as the denominator.
    $\frac{14}{30} = \frac{}{60}$
    30 multiplied by 2 is 60.
    14 multiplied by 2 is 28.
    $\frac{14}{30} = \frac{28}{60}$

Page 21

1. $\frac{1}{2}$ ton
   $\frac{1000}{2000} = \frac{1}{2}$ ton
2. $\frac{7}{20}$ ton
   $\frac{700}{2000} = \frac{7}{20}$ ton
3. $\frac{7}{8}$ pound
   $\frac{14}{16} = \frac{7}{8}$ pound
4. $\frac{1}{2}$ pound
   $\frac{8}{16} = \frac{1}{2}$ pound
5. $\frac{1}{20}$ ton
   $\frac{100}{2000} = \frac{1}{20}$ ton
6. $\frac{3}{4}$ pound
   $\frac{12}{16} = \frac{3}{4}$ pound

Page 22

1. $\frac{3}{8}$ three eighths
2. $\frac{5}{6}$ five sixths
3. $\frac{3}{4}$ three fourths
4. $\frac{7}{10}$ seven tenths
5. $\frac{5}{12}$
6. $\frac{2}{7}$
7. $\frac{13}{24}$
8. $\frac{3}{52}$
9. $\frac{3}{5}$
10. $\frac{3}{7}$
11. $\frac{2}{3}$
12. $\frac{9}{10}$
13. LT
14. $\frac{1}{5}$

15. $\frac{1}{2}$ 16. $\frac{1}{3}$
17. $\frac{1}{2}$ 18. $\frac{1}{8}$
19. $\frac{2}{4}$ 20. $\frac{6}{9}$
21. $\frac{12}{15}$ 22. $\frac{9}{21}$
23. $\frac{25}{30}$ 24. $\frac{12}{16}$
25. $\frac{15}{24}$ 26. $\frac{35}{50}$

Page 23

1. $\frac{3}{8} > \frac{2}{10}$ 2. $\frac{5}{8} < \frac{7}{8}$
3. $\frac{5}{10} = \frac{4}{8}$
4. $\frac{3}{8} < \frac{5}{8}$
   $\frac{5}{8}$ is to the right of $\frac{3}{8}$ on the number line.
5. $\frac{3}{4} > \frac{7}{10}$ 6. $\frac{4}{10} < \frac{5}{8}$
7. $\frac{1}{8} > \frac{1}{10}$ 8. $\frac{3}{8} > \frac{3}{10}$
9. $\frac{1}{2} = \frac{5}{10}$ 10. $\frac{7}{10} < \frac{7}{8}$
11. $\frac{9}{10} > \frac{3}{4}$ 12. $\frac{1}{4} > \frac{1}{10}$

Page 24

1. $\frac{2}{5} = \frac{6}{15}$
2. $\frac{3}{8} < \frac{1}{2}$
   Raise $\frac{1}{2}$ to higher terms with 8 as the denominator.
   $\frac{1}{2} = \frac{1 \times 4}{2 \times 4} = \frac{4}{8}$
   Compare the numerators.
   3 is less than 4.
   $\frac{3}{8} < \frac{4}{8}$, so $\frac{3}{8} < \frac{1}{2}$
3. $\frac{2}{4} > \frac{3}{8}$ 4. $\frac{4}{5} > \frac{2}{10}$
5. $\frac{1}{2} < \frac{6}{10}$ 6. $\frac{4}{9} < \frac{2}{3}$
7. $\frac{1}{4} < \frac{4}{12}$ 8. $\frac{5}{6} > \frac{2}{3}$
9. $\frac{1}{4} < \frac{3}{8}$ 10. $\frac{2}{7} < \frac{5}{14}$
11. $\frac{3}{4} > \frac{7}{16}$ 12. $\frac{4}{7} > \frac{10}{21}$
13. $\frac{5}{8} < \frac{13}{16}$ 14. $\frac{7}{12} > \frac{1}{2}$
15. $\frac{2}{5} < \frac{10}{15}$ 16. $\frac{3}{4} < \frac{16}{20}$
17. $\frac{7}{9} < \frac{15}{18}$ 18. $\frac{3}{10} < \frac{10}{20}$
19. $\frac{5}{8} = \frac{15}{24}$ 20. $\frac{5}{10} = \frac{1}{2}$

21. Pat had completed more work.
    $\frac{2}{3} = \frac{2 \times 3}{3 \times 3} = \frac{6}{9}$
    6 is more than 5.
    $\frac{6}{9} > \frac{5}{9}$, so $\frac{2}{3} > \frac{5}{9}$
22. She has enough.
    $\frac{1}{2} = \frac{1 \times 4}{2 \times 4} = \frac{4}{8}$
    5 is more than 4.
    $\frac{5}{8} > \frac{4}{8}$, so $\frac{5}{8} > \frac{1}{2}$

Page 25

1. $\frac{1}{5} > \frac{1}{6}$
2. $\frac{2}{3} < \frac{3}{4}$
   Multiply the two denominators to find a common denominator. Write each fraction in higher terms with the new denominator.
   $\frac{2}{3} = \frac{2 \times 4}{3 \times 4} = \frac{8}{12}$
   $\frac{3}{4} = \frac{3 \times 3}{4 \times 3} = \frac{9}{12}$
   Compare the numerators.
   $\frac{8}{12} < \frac{9}{12}$, so $\frac{2}{3} < \frac{3}{4}$
3. $\frac{1}{2} < \frac{3}{5}$ 4. $\frac{5}{6} > \frac{4}{7}$
5. $\frac{3}{4} > \frac{3}{5}$ 6. $\frac{4}{5} > \frac{4}{7}$
7. $\frac{3}{8} < \frac{2}{3}$ 8. $\frac{7}{10} > \frac{1}{3}$
9. $\frac{1}{2} = \frac{5}{10}$ 10. $\frac{3}{4} < \frac{9}{10}$
11. $\frac{1}{3} > \frac{1}{4}$ 12. $\frac{2}{4} = \frac{3}{6}$
13. Miami
    $\frac{3}{10} = \frac{3 \times 4}{10 \times 4} = \frac{12}{40}$
    $\frac{3}{4} = \frac{3 \times 10}{4 \times 10} = \frac{30}{40}$
    $\frac{30}{40} > \frac{12}{40}$, so $\frac{3}{4} > \frac{3}{10}$
14. Peterboro
    $\frac{3}{4} = \frac{3}{4}$
    $\frac{1}{2} = \frac{1 \times 2}{1 \times 2} = \frac{2}{4}$
    $\frac{3}{4} > \frac{2}{4}$, so $\frac{3}{4} > \frac{1}{2}$

Page 26

1. $1\frac{1}{2}$
2. $2\frac{1}{3}$
   2 whole figures are shaded. $\frac{1}{3}$ of the remaining figure is shaded.
3. $4\frac{2}{4}$ or $4\frac{1}{2}$ 4. $2\frac{4}{6}$ or $2\frac{2}{3}$

5. $3\frac{5}{8}$
6. $5\frac{1}{2}$
7. $4\frac{1}{8}$ gallons
8. $3\frac{1}{2}$ pies

Page 27
1. $2\frac{1}{2}$
2. $2\frac{1}{4}$
Divide. Write the remainder as a fraction. The remainder, 1, is the numerator of the fraction. The number you divided by, 4, is the denominator.

$$4)\overline{9} \quad R1 = 2\frac{1}{4}$$
$$\underline{-8}$$
$$1$$

3. $1\frac{3}{7}$
4. $2\frac{2}{5}$
5. $6\frac{2}{3}$
6. $2\frac{1}{4}$
7. $8\frac{1}{2}$
8. $2\frac{1}{5}$
9. $2\frac{4}{7}$
10. $2\frac{1}{2}$
11. $2\frac{2}{5}$
12. $6\frac{2}{3}$

Page 28
1. $2\frac{3}{8} > 2\frac{1}{4}$
2. $3\frac{1}{3} < 4\frac{1}{8}$
Compare the whole number parts. 3 is less than 4.

3. $4\frac{1}{5} = 4\frac{2}{10}$
4. $5\frac{3}{4} < 5\frac{7}{8}$
5. $11\frac{1}{2} > 11\frac{1}{10}$
6. $7 < 7\frac{1}{5}$
7. $\frac{1}{5} < 1\frac{1}{5}$
8. $12\frac{3}{4} < 12\frac{15}{16}$
9. $2\frac{1}{2} > 2\frac{1}{12}$
10. $8\frac{7}{8} > 8\frac{9}{16}$
11. $5\frac{1}{3} = 5\frac{5}{15}$
12. $\frac{9}{10} < 10\frac{1}{10}$

Page 29
1. $4\frac{1}{8}$ inches
2. $4\frac{9}{16}$ inches
3. 5 inches
4. $5\frac{1}{2}$ inches

Page 30
5.–16.

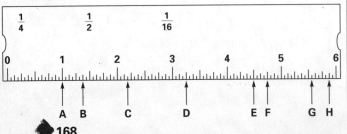

17. $3\frac{3}{8}$ inches
18. $1\frac{1}{2}$ inches
19. $\frac{3}{4}$ inch
20. LT
21. $5\frac{1}{2}$ inches
22. $2\frac{1}{4}$ inches

Page 31
1. $\frac{8}{4}$
2. $\frac{14}{6}$
Count the number of equal parts in each figure. Write this number, 6, as the denominator. Count the number of shaded pieces. Write this number, 14, as the numerator.

3. $\frac{11}{2}$
4. $\frac{11}{3}$
5. $\frac{12}{8}$
6. $\frac{15}{10}$
7. $\frac{18}{12}$
8. $\frac{7}{4}$
9. $\frac{13}{8}$

Page 32
1. M
2. P
The numerator, 2, is smaller than the denominator, 5.

3. I
4. I
5. M
6. W
7. I
8. I
9. W
10. I
11. P
12. P
13. M
14. I
15. W
16. M
17. M
18. I
19. I
20. I
21. $\frac{11}{8}$
One whole melon = $\frac{8}{8}$. Three slices of the second melon = $\frac{3}{8}$. Add the numerators. $\frac{11}{8}$ is an improper fraction.

22. $1\frac{3}{8}$
Walter ate 1 whole melon. He ate 3 of the 8 slices of the second melon. Write 1 for the whole number and $\frac{3}{8}$ for the fraction.

Page 33
1. $\frac{17}{6} = 2\frac{5}{6}$
2. $\frac{9}{2} = 4\frac{1}{2}$
The number of equal parts in each whole, 2, is the denominator. The total number of

shaded parts, 9, is the numerator. Divide the denominator into the numerator to change $\frac{9}{2}$ to a mixed number.

$$2\overline{)9} \quad \begin{array}{r} 4 \\ -8 \\ \hline 1 \end{array} \text{R1} = 4\frac{1}{2}$$

3. $\frac{14}{9} = 1\frac{5}{9}$
4. $\frac{6}{3} = 2$
5. $\frac{6}{5} = 1\frac{1}{5}$
6. $\frac{11}{4} = 2\frac{3}{4}$
7. $2\frac{1}{4}$
8. $1\frac{4}{9}$

$$9\overline{)13} \quad \begin{array}{r} 1 \\ -9 \\ \hline 4 \end{array} \text{R4} = 1\frac{4}{9}$$

9. 2
10. $1\frac{3}{5}$
11. 3
12. $1\frac{5}{8}$

$$\frac{13}{8} = \quad 8\overline{)13} \quad \begin{array}{r} 1 \\ -8 \\ \hline 5 \end{array} \text{R5} = 1\frac{5}{8}$$

13. $2\frac{2}{9}$
14. $2\frac{2}{7}$

Page 34
1. $5\frac{1}{6}$
2. 2

Divide the denominator, 3, into the numerator, 6. 3 goes evenly into 6. The answer is a whole number, 2.

$$3\overline{)6} \quad \begin{array}{r} 2 \\ -6 \\ \hline 0 \end{array}$$

3. 5
4. $8\frac{3}{8}$
5. $2\frac{2}{5}$
6. $1\frac{8}{9}$
7. $6\frac{1}{9}$
8. $9\frac{1}{3}$
9. $2\frac{2}{9}$
10. $3\frac{3}{5}$
11. 5
12. $2\frac{8}{15}$
13. $2\frac{8}{15}$
14. 6
15. $7\frac{1}{9}$
16. $7\frac{4}{11}$
17. $23\frac{1}{3}$
18. 4
19. $5\frac{1}{4}$
20. $1\frac{1}{7}$

21. $4\frac{1}{3}$ pies

The number of pieces in each pie, 6, is the denominator. The total number of pieces sold, 26, is the numerator. Divide to find the number of pies sold. Reduce to lowest terms.

$$\frac{26}{6} = \quad 6\overline{)26} \quad \begin{array}{r} 4 \\ -24 \\ \hline 2 \end{array} \text{R2} = 4\frac{2}{6} = 4\frac{1}{3}$$

22. $3\frac{1}{2}$ dozen

The number of eggs in a dozen, 12, is the denominator. The total number of eggs used, 42, is the numerator. Divide to find how many dozen were used. Reduce to lowest terms.

$$12\overline{)42} \quad \begin{array}{r} 3 \\ -36 \\ \hline 6 \end{array} \text{R6} = 3\frac{6}{12} = 3\frac{1}{2}$$

Page 35
1. $1\frac{1}{2} = \frac{3}{2}$
2. $2\frac{2}{3} = \frac{8}{3}$

2 whole circles are shaded. 2 of the 3 parts in the third circle are shaded. Write as a mixed number, $2\frac{2}{3}$. Multiply the denominator, 3, by the whole number, 2. $3 \times 2 = 6$. Add 6 to the numerator, 2. $6 + 2 = 8$. Write 8 over 3. $\frac{8}{3}$ is an improper fraction.

3. $2\frac{2}{6} = \frac{14}{6}$
4. $1\frac{4}{8} = \frac{12}{8}$
5. $1\frac{2}{5} = \frac{7}{5}$
6. $3\frac{3}{4} = \frac{15}{4}$
7. $4\frac{2}{4} = \frac{18}{4}$
8. $1\frac{1}{4} = \frac{5}{4}$
9. $2\frac{3}{8} = \frac{19}{8}$

Page 36
1. $\frac{10}{3}$
2. $\frac{32}{5}$

Write the denominator, 5, for the improper fraction. Multiply the denominator, 5, by the whole number, 6. $5 \times 6 = 30$. Add the numerator of the fraction, 2, to the answer 30. $30 + 2 = 32$. Write the sum over the denominator, 5.

3. $\frac{105}{8}$
4. $\frac{98}{8}$

169

5. $\frac{15}{2}$ 6. $\frac{26}{3}$
7. $\frac{107}{10}$ 8. $\frac{59}{6}$
9. $\frac{36}{5}$ 10. $\frac{111}{7}$
11. $\frac{13}{12}$ 12. $\frac{37}{9}$
13. $\frac{10}{2}$
Write the length of each piece, 2, as the denominator. Write the total length of the plank, 10, as the numerator.
14. $\frac{9}{2}$
$4\frac{1}{2} = \frac{9}{2}$
$2 \times 4 = 8$
$8 + 1 = 9$

Unit 1 Review, page 37
1. $\frac{1}{3}$ 2. $\frac{2}{5}$
3. $\frac{5}{6}$ 4. $\frac{7}{4} = 1\frac{3}{4}$
5. $\frac{9}{3} = 3$ 6. $\frac{13}{6} = 2\frac{1}{6}$
7. $\frac{15}{6} = 2\frac{3}{6} = 2\frac{1}{2}$

8.–10. Shading should be similar to this.

8.   9.

10.

11. $\frac{3}{4}$ 12. LT
13. $\frac{3}{7}$ 14. $\frac{2}{3}$
15. $\frac{4}{7}$ 16. $\frac{3}{4}$
17. LT 18. $\frac{2}{3}$
19. $\frac{2}{3}$ 20. $\frac{1}{2}$

Page 38
21. $\frac{4}{6}$ 22. $\frac{3}{24}$
23. $\frac{4}{18}$ 24. $\frac{20}{28}$
25. $\frac{5}{10}$ 26. $\frac{9}{15}$
27. $\frac{15}{20}$ 28. $\frac{27}{30}$

29. $\frac{1}{2} > \frac{2}{5}$ 30. $\frac{2}{3} < \frac{15}{20}$
31. $\frac{4}{7} < \frac{7}{10}$ 32. $4\frac{1}{9} < 4\frac{1}{3}$
33. $\frac{5}{8} = \frac{10}{16}$ 34. $7\frac{3}{9} > 7\frac{1}{4}$
35. $\frac{4}{10} < \frac{2}{4}$ 36. $\frac{6}{8} = \frac{3}{4}$
37. $1\frac{1}{2}$ 38. $3\frac{3}{4}$
39. 4 40. $1\frac{1}{2}$
41. $6\frac{3}{7}$ 42. $8\frac{7}{8}$
43. $3\frac{2}{9}$ 44. 4
45. $\frac{25}{6}$ 46. $\frac{27}{5}$
47. $\frac{40}{3}$ 48. $\frac{13}{8}$

Unit 2

Page 39
1. $\frac{1}{3} < \frac{2}{3}$
2. $\frac{7}{10} > \frac{3}{5}$
Change $\frac{3}{5}$ to an equal fraction with a denominator of 10.
$\frac{3}{5} = \frac{3 \times 2}{5 \times 2} = \frac{6}{10}$
Compare the numerators. 7 is larger than 6.
$\frac{7}{10} > \frac{6}{10}$, so $\frac{7}{10} > \frac{3}{5}$
3. $\frac{9}{12} < \frac{5}{6}$ 4. $\frac{2}{3} < \frac{3}{4}$
5. $\frac{2}{9} < \frac{1}{3}$ 6. $\frac{3}{8} = \frac{6}{16}$
7. $\frac{8}{11} > \frac{5}{11}$ 8. $\frac{1}{2} < \frac{2}{3}$

Page 40
9. $\frac{2}{3}$
10. $\frac{3}{4}$
$\frac{15}{20} = \frac{15 \div 5}{20 \div 5} = \frac{3}{4}$
11. $\frac{1}{2}$ 12. $3\frac{3}{4}$
13. $1\frac{2}{3}$ 14. $\frac{2}{3}$
15. $\frac{4}{5}$ 16. $4\frac{1}{2}$
17. $5\frac{1}{2}$ 18. $10\frac{3}{4}$
19. $\frac{8}{12}$
20. $\frac{9}{12}$
$\frac{3}{4} = \frac{3 \times 3}{4 \times 3} = \frac{9}{12}$

21. $\frac{6}{12}$ **22.** $2\frac{10}{12}$
23. $5\frac{3}{12}$ **24.** $\frac{10}{12}$
25. $\frac{4}{12}$ **26.** $9\frac{6}{12}$
27. $3\frac{8}{12}$ **28.** $1\frac{8}{12}$
29. $2\frac{1}{5}$
30. $6\frac{2}{3}$

$$3\overline{)20} \quad R2 = 6\frac{2}{3}$$
$$\underline{-18}$$
$$2$$

31. $5\frac{4}{9}$ **32.** 3
33. 5 **34.** $3\frac{1}{2}$
35. $1\frac{3}{5}$ **36.** 2
37. $2\frac{1}{2}$ **38.** $1\frac{1}{4}$

Page 41

1. $\frac{2}{3}$

2.–4. and 6. Shading should be similar to this.

2. $\frac{3}{4}$

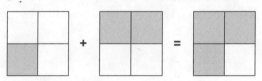

Add the numerators.
$1 + 3 = 4$. Write the sum over the denominator, 4. Shade in 3 of the 4 squares in the final figure.

3. $\frac{4}{5}$

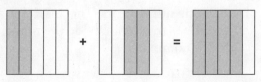

4. $\frac{5}{6}$

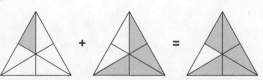

5.

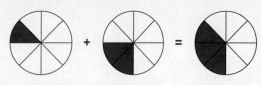

$\frac{3}{8}$

6.

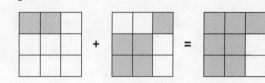

$\frac{7}{9}$

Shade in 2 squares in the first figure. Shade in 5 squares in the second figure. Add $2 + 5$. The sum, 7, is the numerator over the denominator, 9. Shade in 7 of the 9 squares in the final figure.

Page 42

1. $\frac{5}{6}$

2. $\frac{7}{12}$

The denominators are the same, 12. Add the numerators. $6 + 1 = 7$. Write the sum over the denominator.

$\frac{6}{12} + \frac{1}{12} = \frac{7}{12}$

3. $\frac{13}{15}$ **4.** $\frac{11}{20}$
5. $\frac{9}{10}$ **6.** $\frac{13}{14}$
7. $\frac{31}{35}$

The denominators are the same, 35. Add the numerators. $10 + 15 + 6 = 31$. Write the sum over the denominator.

$\frac{10}{35} + \frac{15}{35} + \frac{6}{35} = \frac{31}{35}$

8. $\frac{21}{40}$ **9.** $\frac{31}{50}$
10. $\frac{33}{100}$ **11.** $\frac{14}{25}$
12. $\frac{7}{9}$ **13.** $\frac{5}{7}$
14. $\frac{4}{5}$

Add $\frac{2}{5} + \frac{2}{5}$. The denominators are the same, 5. Add the numerators. $2 + 2 = 4$. Write the sum over the denominator.

$\frac{2}{5} + \frac{2}{5} = \frac{4}{5}$

15. $\frac{5}{8}$ gallon

Add $\frac{2}{8}+\frac{3}{8}$. The denominators are the same, 8. Add the numerators. $2+3=5$. Write the sum over the denominator.
$\frac{2}{8}+\frac{3}{8}=\frac{5}{8}$

Page 43

1. $\frac{1}{2}$ hour

   Ann spent $\frac{1}{4}$ hour on the rowing machine. Otis spent $\frac{1}{4}$ hour on the rowing machine. Petra did not spend any time on the rowing machine. Add $\frac{1}{4}+\frac{1}{4}$. Reduce to lowest terms.
   $\frac{1}{4}+\frac{1}{4}=\frac{2}{4}=\frac{2\div 2}{4\div 2}=\frac{1}{2}$

2. 1 hour

   Ann spent $\frac{2}{4}$ hour on the exercise bike, $\frac{1}{4}$ hour on the treadmill, and $\frac{1}{4}$ hour on the rowing machine. Add $\frac{2}{4}+\frac{1}{4}+\frac{1}{4}$. Reduce to lowest terms.
   $\frac{2}{4}+\frac{1}{4}+\frac{1}{4}=\frac{4}{4}=1$

3. $1\frac{1}{2}$ hours

   Otis spent $\frac{3}{4}$ hour on the exercise bike, $\frac{2}{4}$ hour on the treadmill, and $\frac{1}{4}$ hour on the rowing machine. Add $\frac{3}{4}+\frac{2}{4}+\frac{1}{4}$. Reduce to lowest terms.
   $\frac{3}{4}+\frac{2}{4}+\frac{1}{4}=\frac{6}{4}=1\frac{2}{4}=1\frac{1}{2}$

4. $\frac{3}{4}$ hour

   Otis spent $\frac{2}{4}$ hour on the treadmill. Ann spent $\frac{1}{4}$ hour. Add $\frac{2}{4}+\frac{1}{4}$.
   $\frac{2}{4}+\frac{1}{4}=\frac{3}{4}$

Page 44

1. $\frac{1}{2}$
2. $\frac{2}{3}$

   The denominators are the same, 6. Add the numerators. $3+1=4$. Write the sum over the denominator. Reduce to lowest terms.
   $\frac{4}{6}=\frac{4\div 2}{6\div 2}=\frac{2}{3}$

3. $\frac{4}{5}$
4. $\frac{2}{3}$
5. $\frac{2}{3}$
6. $\frac{3}{5}$
7. $\frac{2}{3}$
8. $\frac{1}{4}$
9. $\frac{17}{22}$
10. $\frac{7}{8}$
11. $\frac{9}{10}$
12. $\frac{7}{9}$
13. $\frac{5}{6}$
14. $\frac{1}{3}$
15. $\frac{2}{3}$
16. $\frac{5}{6}$
17. $\frac{4}{5}$
18. $\frac{3}{4}$
19. $\frac{7}{10}$
20. $\frac{4}{5}$ mile

    Add how far it is from the stop sign to the gas station, $\frac{2}{10}$, to how far it is from the gas station to the traffic signal, $\frac{6}{10}$. Reduce to lowest terms.
    $\frac{2}{10}+\frac{6}{10}=\frac{8}{10}=\frac{8\div 2}{10\div 2}=\frac{4}{5}$

21. $\frac{1}{2}$ inch

    Add the widths of the two ribbons. Reduce to lowest terms.
    $\frac{2}{16}+\frac{6}{16}=\frac{8}{16}=\frac{8\div 8}{16\div 8}=\frac{1}{2}$

Page 45

1. 1
2. 1

   The denominators are the same. Add the numerators. The sum is an improper fraction. Divide the numerator by the denominator.
   $\frac{5}{6}+\frac{1}{6}=\frac{6}{6}$
   $$6\overline{)6}\begin{array}{r}1\\-6\\\hline 0\end{array}$$

3. 1
4. 1
5. 3
6. 2

   Add the numerators. The answer is an improper fraction. Divide the numerator by the denominator.
   $\frac{6}{4}+\frac{2}{4}=\frac{8}{4}$
   $$4\overline{)8}\begin{array}{r}2\\-8\\\hline 0\end{array}$$

7. 3
8. 2
9. 2
10. 3

11. 3
12. 1
13. 3
14. 4
15. 1
16. 5
17. 2 feet
    Add the 3 lengths of pipe.
    $\frac{9}{12} + \frac{9}{12} + \frac{6}{12} = \frac{24}{12}$
    Change the improper fraction to a whole number.
    $12 \overline{)24}$
    $\underline{-24}$
    0 → 2
18. 1 inch
    Add the rainfall for all 3 days last week.
    $\frac{2}{10} + \frac{1}{10} + \frac{7}{10} = \frac{10}{10}$
    Change the improper fraction to a whole number.
    $10 \overline{)10}$
    $\underline{-10}$
    0 → 1

Page 46

1. $1\frac{1}{3}$
2. $3\frac{1}{2}$
   The denominators are the same. Add the numerators. The answer is an improper fraction. Change to a mixed number. Reduce.
   $\frac{6}{4} + \frac{8}{4} = \frac{14}{4}$
   $4\overline{)14} = 3\frac{2}{4} = 3\frac{1}{2}$
   $\underline{-12}$
   2
3. $1\frac{2}{5}$
4. $1\frac{2}{3}$
5. $2\frac{1}{5}$
6. $4\frac{1}{3}$
   The denominators are the same. Add the numerators. The answer is an improper fraction. Change to a mixed number.
   $\frac{11}{3} + \frac{1}{3} + \frac{1}{3} = \frac{13}{3}$
   $3\overline{)13} = 4\frac{1}{3}$
   $\underline{-12}$
   1
7. $1\frac{8}{9}$
8. $2\frac{1}{2}$

9. $1\frac{8}{15}$
10. $1\frac{3}{20}$
11. $1\frac{3}{7}$
12. $1\frac{1}{5}$
13. $1\frac{1}{2}$ hours
    Add the amounts of time Anita spent in the park. Change to a mixed number. Reduce.
    $\frac{3}{4} + \frac{3}{4} = \frac{6}{4}$
    $4\overline{)6} = 1\frac{2}{4} = 1\frac{1}{2}$
    $\underline{-4}$
    2
14. $1\frac{7}{10}$ miles
    Add the distance from Adams to Bailey, $\frac{8}{10}$, to the distance from Bailey to Carter, $\frac{9}{10}$. Change to a mixed number.
    $\frac{8}{10} + \frac{9}{10} = \frac{17}{10}$
    $10\overline{)17} = 1\frac{7}{10}$
    $\underline{-10}$
    7

Page 47

1. $\frac{1}{2}$
2. 3
3. $4\frac{1}{3}$
4. $\frac{1}{4}$
5. $3\frac{3}{4}$
6. $12\frac{1}{2}$
7. $3\frac{1}{4}$
8. 1
9. $\frac{1}{7}$
10. LT
11. $1\frac{4}{7}$
12. $1\frac{8}{11}$
13. $1\frac{1}{5}$
14. $1\frac{1}{6}$
15. $\frac{2}{5}$
16. $\frac{9}{10}$
17. $\frac{22}{25}$
18. $\frac{71}{100}$
19. $\frac{4}{5}$
20. $\frac{2}{3}$
21. $\frac{5}{6}$
22. $\frac{4}{5}$
23. 1
24. 2
25. 3
26. 1
27. $1\frac{7}{9}$
28. $1\frac{1}{3}$
29. $2\frac{11}{25}$
30. $1\frac{1}{6}$
31. $\frac{7}{8}$
32. $1\frac{5}{7}$

33. $\frac{2}{3}$ **34.** $1\frac{2}{5}$

35. $1\frac{1}{3}$

Page 48

1. 3 sections were used for football.
   $\frac{3}{12} = \frac{1}{4}$

2. 7 sections; $\frac{7}{12}$

3. 1 section; $\frac{1}{12}$

4. $\frac{2}{12} = \frac{1}{6}$ **5.** $\frac{9}{12} = \frac{3}{4}$

6. 1
   $\frac{3}{12} + \frac{4}{12} + \frac{2}{12} + \frac{2}{12} + \frac{1}{12} = \frac{12}{12} = 1$

Page 49

1. $9\frac{1}{2}$

2. $8\frac{1}{2}$
   Add the fractions. Add the whole numbers. Reduce.

   $1\frac{1}{4}$
   $+ 7\frac{1}{4}$
   $\overline{ 8\frac{2}{4}} = 8\frac{1}{2}$

3. $7\frac{7}{8}$ **4.** $9\frac{4}{5}$

5. $21\frac{19}{20}$ **6.** $41\frac{4}{5}$

7. $63\frac{7}{10}$ **8.** $79\frac{3}{5}$

9. $17\frac{1}{2}$

10. $45\frac{5}{7}$
    Line up the fractions and the whole numbers in columns. Add the fractions. Add the whole numbers.

    $10\frac{2}{7}$
    $13\frac{1}{7}$
    $+ 22\frac{2}{7}$
    $\overline{ 45\frac{5}{7}}$

11. $16\frac{1}{2}$

Page 50

1. 7

2. $11\frac{1}{2}$
   Line up the mixed numbers. Add the fractions. Add the whole numbers.

   $5\frac{3}{4}$
   $+ 5\frac{3}{4}$
   $\overline{ 10\frac{6}{4}}$

   Change $\frac{6}{4}$ to a mixed number, $1\frac{2}{4}$. Add the whole numbers. Reduce the fraction.
   $10\frac{6}{4} = 10 + 1\frac{2}{4} = 11\frac{2}{4} = 11\frac{1}{2}$

3. $25\frac{2}{5}$ **4.** $27\frac{1}{9}$

5. 41   **6.** $110\frac{3}{10}$

7. 159   **8.** 46

9. $5\frac{1}{2}$ hours
   Add the amounts of time Phil watched TV on the 3 days given. Change the fraction to a mixed number. Add it to the whole number.

   $1\frac{1}{2}$
   $2\frac{1}{2}$
   $+ 1\frac{1}{2}$
   $\overline{ 4\frac{3}{2}} = 4 + 1\frac{1}{2} = 5\frac{1}{2}$

10. $10\frac{1}{2}$ miles
    Add all the distances. Change the answer to a mixed number.

    $3\frac{7}{10}$
    $1\frac{3}{10}$
    $+ 6\frac{1}{10}$
    $\overline{ 10\frac{11}{10}} = 10 + 1\frac{1}{10} = 11\frac{1}{10}$

Page 51

1. $7\frac{1}{4}$

2. $11\frac{1}{5}$

   Add the fractions. Add the whole numbers. Change the improper fraction to a mixed number. Add it to the whole number.

   $$\begin{array}{r} 3\frac{2}{5} \\ 7 \\ +\ \frac{4}{5} \\ \hline 10\frac{6}{5} = 10 + 1\frac{1}{5} = 11\frac{1}{5} \end{array}$$

3. $16\frac{3}{10}$ 4. 16
5. $34\frac{5}{9}$ 6. $18\frac{1}{4}$
7. $7\frac{3}{4}$ yards

   Add the amounts of material given. Line up the whole numbers and fractions. Add the fractions. Add the whole numbers. Reduce to lowest terms.

   $$\begin{array}{r} 3\frac{1}{8} \\ 4 \\ +\ \frac{5}{8} \\ \hline 7\frac{6}{8} = 7\frac{3}{4} \end{array}$$

8. $4\frac{3}{4}$ yards

   Add the amounts of material given. Add the fractions. Add the whole numbers. Reduce to lowest terms.

   $$\begin{array}{r} 2\frac{3}{8} \\ 2 \\ +\ \frac{3}{8} \\ \hline 4\frac{6}{8} = 4\frac{3}{4} \end{array}$$

Page 52

1. LT   2. $4\frac{1}{2}$
3. 5   4. LT
5. LT   6. $\frac{3}{4}$
7. $8\frac{1}{4}$ 8. LT
9. 3   10. 1
11. $\frac{5}{9}$ 12. $\frac{3}{4}$
13. $1\frac{2}{5}$ 14. $1\frac{5}{14}$
15. $12\frac{1}{4}$ 16. 10
17. $11\frac{1}{4}$ 18. $33\frac{1}{2}$
19. $13\frac{7}{8}$ 20. $17\frac{1}{2}$
21. $1\frac{3}{4}$ 22. $25\frac{5}{8}$

23. $72\frac{4}{25}$
24. $63\frac{1}{2}$ inches

    $$\begin{array}{r} 62\frac{3}{4} \\ +\ \frac{3}{4} \\ \hline 62\frac{6}{4} = 62 + 1\frac{2}{4} = 63\frac{2}{4} = 63\frac{1}{2} \end{array}$$

25. 64 inches

    $$\begin{array}{r} 62\frac{3}{4} \\ +\ 1\frac{1}{4} \\ \hline 63\frac{4}{4} = 63 + 1 = 64 \end{array}$$

Page 53

1. $1\frac{3}{4}$ inches

   Add the measures. Draw an arrow pointing to the answer and label it B.

   $$\begin{array}{r} 1\frac{1}{2} = 1\frac{2}{4} \\ +\ \frac{1}{4} = \frac{1}{4} \\ \hline 1\frac{3}{4} \end{array}$$

2. $3\frac{5}{8}$ inches   3. $5\frac{3}{4}$ inches
4. $5\frac{9}{16}$ inches

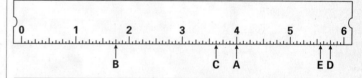

Page 54

5. $3\frac{3}{4}$ inches   6. $4\frac{1}{8}$ inches
7. $3\frac{1}{2}$ inches   8. $4\frac{9}{16}$ inches
9. $2\frac{3}{4}$ inches   10. $5\frac{1}{4}$ inches
11. $4\frac{3}{8}$ inches 12. 6 inches

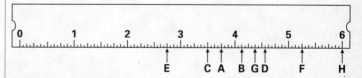

Page 55

1. $\frac{1}{2} = \frac{5}{10}$

   $\frac{3}{10} = \frac{3}{10}$

2. $\frac{4}{5} = \frac{12}{15}$
$\frac{7}{15} = \frac{7}{15}$

Compare the denominators. 5 divides evenly into 15. Raise $\frac{4}{5}$ to higher terms with 15 as the denominator.
$\frac{4}{5} = \frac{4 \times 3}{5 \times 3} = \frac{12}{15}$

3. $\frac{2}{3} = \frac{6}{9}$
$\frac{4}{9} = \frac{4}{9}$

4. $\frac{7}{12} = \frac{7}{12}$
$\frac{5}{6} = \frac{10}{12}$

5. $\frac{3}{4} = \frac{6}{8}$
$\frac{5}{8} = \frac{5}{8}$

6. $\frac{1}{2} = \frac{3}{6}$
$\frac{5}{6} = \frac{5}{6}$

7. $\frac{7}{21} = \frac{7}{21}$
$\frac{3}{7} = \frac{9}{21}$

8. $\frac{1}{5} = \frac{5}{25}$
$\frac{3}{25} = \frac{3}{25}$

9. $\frac{3}{14} = \frac{3}{14}$
$\frac{2}{7} = \frac{4}{14}$

10. $\frac{5}{8} = \frac{15}{24}$
$\frac{7}{24} = \frac{7}{24}$

11. $\frac{9}{20} = \frac{9}{20}$
$\frac{3}{10} = \frac{6}{20}$

12. $\frac{2}{9} = \frac{6}{27}$
$\frac{5}{27} = \frac{5}{27}$

13. $\frac{1}{2} = \frac{4}{8}$
$\frac{3}{4} = \frac{6}{8}$
$\frac{7}{8} = \frac{7}{8}$

14. $\frac{2}{3} = \frac{12}{18}$
$\frac{5}{6} = \frac{15}{18}$
$\frac{1}{18} = \frac{1}{18}$

Compare the denominators. The two smaller denominators, 3 and 6, will divide evenly into the largest denominator, 18. Raise $\frac{2}{3}$ and $\frac{5}{6}$ to higher terms with 18 as the denominator.
$\frac{2}{3} = \frac{2 \times 6}{3 \times 6} = \frac{12}{18}$
$\frac{5}{6} = \frac{5 \times 3}{6 \times 3} = \frac{15}{18}$

15. $\frac{1}{3} = \frac{5}{15}$
$\frac{3}{5} = \frac{9}{15}$
$\frac{7}{15} = \frac{7}{15}$

Page 56

1. $\frac{7}{10}$

2. $\frac{5}{6}$

Compare the denominators. 2 divides evenly into 6. Raise $\frac{1}{2}$ to higher terms with 6 as the denominator.
$\frac{1}{2} = \frac{1 \times 3}{2 \times 3} = \frac{3}{6}$

Add the fractions.
$\frac{2}{6} = \frac{2}{6}$
$+\frac{1}{2} = \frac{3}{6}$
$\overline{\frac{5}{6}}$

3. $\frac{5}{14}$
4. $\frac{3}{4}$
5. $\frac{7}{8}$
6. $\frac{8}{9}$
7. $\frac{7}{10}$
8. $\frac{4}{15}$
9. $\frac{7}{20}$

Page 57

1. $1\frac{1}{5}$

2. $\frac{9}{16}$

Compare the denominators. The two smaller denominators, 4 and 8, will divide evenly into the largest denominator, 16. Raise $\frac{1}{4}$ and $\frac{1}{8}$ to higher terms with 16 as the denominator.
$\frac{1}{4} = \frac{1 \times 4}{4 \times 4} = \frac{4}{16}$
$\frac{1}{8} = \frac{1 \times 2}{8 \times 2} = \frac{2}{16}$

Add the fractions.
$\frac{1}{4} = \frac{4}{16}$
$\frac{1}{8} = \frac{2}{16}$
$+\frac{3}{16} = \frac{3}{16}$
$\overline{\frac{9}{16}}$

3. $1\frac{5}{18}$
4. $\frac{13}{18}$
5. $\frac{13}{16}$
6. $\frac{9}{10}$
7. $\frac{19}{20}$
8. $\frac{19}{45}$
9. $\frac{41}{60}$
10. $\frac{27}{52}$
11. $\frac{99}{100}$

Page 58

1. $\frac{1}{2}$

2. $\frac{1}{3}$

Raise $\frac{1}{5}$ to higher terms with 15 as the denominator.

$\frac{1}{5} = \frac{1 \times 3}{5 \times 3} = \frac{3}{15}$

Add the fractions. Reduce to lowest terms.

$\frac{2}{15} = \frac{2}{15}$
$+\frac{1}{5} = \frac{3}{15}$
$\frac{5}{15} = \frac{1}{3}$

3. $\frac{2}{5}$ **4.** $\frac{24}{25}$
5. $\frac{1}{2}$ **6.** $\frac{5}{6}$
7. $\frac{4}{5}$ **8.** $\frac{33}{50}$
9. $\frac{1}{4}$ **10.** $\frac{7}{8}$
11. $\frac{9}{10}$ **12.** $\frac{5}{6}$

13. $\frac{3}{4}$ cup

Add the amounts of sugar Joan used. Raise $\frac{1}{2}$ to higher terms with 4 as the denominator.

$\frac{1}{2} = \frac{1 \times 2}{2 \times 2} = \frac{2}{4}$

Add the fractions.

$\frac{1}{2} = \frac{2}{4}$
$+\frac{1}{4} = \frac{1}{4}$
$\frac{3}{4}$

14. $\frac{1}{2}$

Add the parts of his allowance that Chris spent. Raise $\frac{1}{3}$ to higher terms with 6 as the denominator.

$\frac{1}{3} = \frac{1 \times 2}{3 \times 2} = \frac{2}{6}$

Add the fractions. Reduce to lowest terms.

$\frac{1}{6} = \frac{1}{6}$
$+\frac{1}{3} = \frac{2}{6}$
$\frac{3}{6} = \frac{1}{2}$

Page 59

1. $1\frac{1}{6}$

2. $1\frac{5}{8}$

Raise $\frac{3}{4}$ to higher terms with 8 as the denominator.

$\frac{3}{4} = \frac{3 \times 2}{4 \times 2} = \frac{6}{8}$

Add the fractions. Change the improper fraction to a mixed number.

$\frac{7}{8} = \frac{7}{8}$
$+\frac{3}{4} = \frac{6}{8}$
$\frac{13}{8} = 1\frac{5}{8}$

3. $1\frac{1}{3}$ **4.** $1\frac{5}{16}$
5. $1\frac{1}{6}$ **6.** 2
7. $1\frac{2}{3}$ **8.** $1\frac{3}{8}$
9. $1\frac{5}{14}$ **10.** $1\frac{7}{12}$
11. $1\frac{8}{15}$ **12.** 1

13. $2\frac{1}{4}$ pounds

Add the weights of the three boxes to find the total weight.

$\frac{12}{16} = \frac{12}{16}$
$\frac{3}{4} = \frac{12}{16}$
$+\frac{6}{8} = \frac{12}{16}$
$\frac{36}{16} = 2\frac{4}{16} = 2\frac{1}{4}$

14. $1\frac{3}{5}$ yards

Add the three lengths of ribbon to find the combined length.

$\frac{4}{5} = \frac{8}{10}$
$\frac{3}{10} = \frac{3}{10}$
$+\frac{1}{2} = \frac{5}{10}$
$\frac{16}{10} = 1\frac{6}{10} = 1\frac{3}{5}$

Page 60

1. $\frac{2}{8}$ **2.** $\frac{9}{12}$
3. $\frac{6}{9}$ **4.** $\frac{15}{18}$
5. $\frac{12}{20}$ **6.** $\frac{2}{3}$
7. $\frac{3}{5}$ **8.** $\frac{1}{6}$
9. LT   **10.** $\frac{1}{2}$
11. 1   **12.** $1\frac{1}{4}$
13. 3   **14.** $3\frac{1}{7}$
15. $2\frac{1}{4}$ **16.** $\frac{1}{2}$
17. 1   **18.** $1\frac{4}{9}$
19. $\frac{1}{2}$ **20.** $\frac{7}{16}$

21. $\frac{5}{8}$ 22. $1\frac{9}{14}$

23. $1\frac{7}{12}$ 24. $2\frac{1}{18}$

25. $\frac{3}{5}$ 26. $5\frac{6}{7}$

27. $14\frac{5}{6}$ 28. $\frac{21}{25}$

29. $10\frac{1}{2}$

Page 61

1. $2\frac{1}{8}$ pounds

   Add the amounts of weight Theresa lost. Raise $\frac{1}{2}$ and $\frac{3}{4}$ to higher terms with 16 as the denominator. Add the fractions. Change the improper fraction to a mixed number. Reduce.

   $\frac{1}{2} = \frac{8}{16}$
   $\frac{3}{4} = \frac{12}{16}$
   $\frac{9}{16} = \frac{9}{16}$
   $+\frac{5}{16} = \frac{5}{16}$
   $\frac{34}{16} = 2\frac{2}{16} = 2\frac{1}{8}$

2. $2\frac{9}{16}$ pounds

   Add the amounts of weight Lisa lost.

   $\frac{3}{4} = \frac{12}{16}$
   $\frac{15}{16} = \frac{15}{16}$
   $\frac{1}{2} = \frac{8}{16}$
   $+\frac{3}{8} = \frac{6}{16}$
   $\frac{41}{16} = 2\frac{9}{16}$

3. $1\frac{11}{16}$ pounds

   Add the amounts of weight Janet lost.

   $\frac{3}{8} = \frac{6}{16}$
   $\frac{1}{4} = \frac{4}{16}$
   $\frac{7}{8} = \frac{14}{16}$
   $+\frac{3}{16} = \frac{3}{16}$
   $\frac{27}{16} = 1\frac{11}{16}$

4. Mona and Lisa lost the same amount of weight, $2\frac{9}{16}$ pounds, so they tied.

Page 62

1. $\frac{1}{5} = \frac{3}{15}$
   $\frac{2}{3} = \frac{10}{15}$

2. $\frac{2}{5} = \frac{8}{20}$
   $\frac{1}{4} = \frac{5}{20}$

   Compare the denominators. 4 will not divide evenly into 5. Multiply the denominators of the two fractions to get the common denominator, 20. Raise each fraction to higher terms with 20 as the denominator.

   $\frac{2}{5} = \frac{2 \times 4}{5 \times 4} = \frac{8}{20}$
   $\frac{1}{4} = \frac{1 \times 5}{4 \times 5} = \frac{5}{20}$

3. $\frac{1}{6} = \frac{5}{30}$ 4. $\frac{3}{7} = \frac{6}{14}$
   $\frac{3}{5} = \frac{18}{30}$ $\frac{1}{2} = \frac{7}{14}$

5. $\frac{1}{3} = \frac{4}{12}$
   $\frac{1}{4} = \frac{3}{12}$
   $\frac{1}{6} = \frac{2}{12}$

6. $\frac{1}{2} = \frac{35}{70}$
   $\frac{2}{5} = \frac{28}{70}$
   $\frac{3}{7} = \frac{30}{70}$

   Compare the two larger denominators. 5 will not divide evenly into 7. Multiply 5×7. $5 \times 7 = 35$. The third denominator, 2, will not divide evenly into 35. Multiply 35×2. $35 \times 2 = 70$. The common denominator is 70. Raise all three fractions to higher terms with 70 as the denominator.

7. $\frac{4}{9} = \frac{28}{63}$ 8. $\frac{5}{9} = \frac{20}{36}$
   $\frac{1}{3} = \frac{21}{63}$ $\frac{1}{4} = \frac{9}{36}$
   $\frac{2}{7} = \frac{18}{63}$

9. $\frac{2}{3} = \frac{28}{42}$ 10. $\frac{3}{4} = \frac{27}{36}$
   $\frac{2}{7} = \frac{12}{42}$ $\frac{2}{9} = \frac{8}{36}$
   $\frac{5}{6} = \frac{35}{42}$

11. $\frac{3}{8} = \frac{45}{120}$

$\frac{3}{5} = \frac{72}{120}$

$\frac{2}{3} = \frac{80}{120}$

Page 63

1. $\frac{5}{6}$

2. $\frac{7}{12}$

Multiply the denominators to find the common denominator, 24. Raise each fraction to higher terms with 24 as the denominator. Add the fractions. Reduce.

$\frac{2}{6} = \frac{8}{24}$

$+\frac{1}{4} = \frac{6}{24}$

$\frac{14}{24} = \frac{7}{12}$

3. $\frac{17}{24}$ **4.** $\frac{15}{28}$

5. $\frac{13}{20}$

6. $\frac{5}{6}$

Multiply the denominators to find the common denominator, 18. Raise each fraction to higher terms with 18 as the denominator. Add the fractions. Reduce.

$\frac{2}{3} = \frac{12}{18}$

$+\frac{1}{6} = \frac{3}{18}$

$\frac{15}{18} = \frac{15 \div 3}{18 \div 3} = \frac{5}{6}$

7. $\frac{5}{9}$ **8.** $\frac{8}{15}$

9. $\frac{11}{12}$ **10.** $\frac{9}{14}$

11. $\frac{33}{40}$ **12.** $\frac{19}{30}$

Page 64

1. $1\frac{13}{24}$

2. $1\frac{11}{20}$

Multiply the two denominators to find the common denominator, 20. Raise each fraction to higher terms with 20 as the denominator. Add the fractions. Change the improper fraction to a mixed number.

$\frac{4}{5} = \frac{16}{20}$

$+\frac{3}{4} = \frac{15}{20}$

$\frac{31}{20} = 1\frac{11}{20}$

3. $1\frac{7}{30}$ **4.** $1\frac{5}{42}$

5. $1\frac{9}{20}$

6. $2\frac{7}{24}$

Multiply 3 × 8 to find a common denominator, 24. 4 will divide evenly into 24. Raise each fraction to higher terms. Add. Change the improper fraction to a mixed number.

$\frac{7}{8} = \frac{21}{24}$

$\frac{3}{4} = \frac{18}{24}$

$+\frac{2}{3} = \frac{16}{24}$

$\frac{55}{24} = 2\frac{7}{24}$

7. $1\frac{19}{140}$

8. $2\frac{1}{12}$ pounds

Add the amounts of the three purchases. Multiply the two larger denominators to find the common denominator, 24. Raise each fraction to higher terms. Add. Reduce to lowest terms.

$\frac{5}{6} = \frac{20}{24}$

$\frac{3}{4} = \frac{18}{24}$

$+\frac{1}{2} = \frac{12}{24}$

$\frac{50}{24} = 2\frac{2}{24} = 2\frac{1}{12}$

9. $1\frac{11}{12}$ pounds

Add the amounts of the three purchases.

$\frac{2}{3} = \frac{8}{12}$

$\frac{3}{4} = \frac{9}{12}$

$+\frac{1}{2} = \frac{6}{12}$

$\frac{23}{12} = 1\frac{11}{12}$

Page 65

1. $\frac{5}{18}$

2. $1\frac{1}{12}$

List the multiples of each denominator. The smallest number on both lists is 12. 12 is the LCD. Raise each fraction to higher terms with 12 as the LCD. Add the fractions. Change the improper fraction to a mixed number.

$\frac{5}{6}$ 6   ⑫   18
$\frac{1}{4}$ 4   8   ⑫

$$\frac{5}{6} = \frac{10}{12}$$
$$+\frac{1}{4} = \frac{3}{12}$$
$$\frac{13}{12} = 1\frac{1}{12}$$

3. $\frac{37}{40}$

$\frac{5}{8}$ 8   16   32   ㊵
$\frac{3}{10}$ 10   20   30   ㊵

4. $\frac{13}{48}$

$\frac{1}{12}$ 12   24   36   ㊽
$\frac{3}{16}$ 16   32   ㊽

5. $\frac{13}{15}$

$\frac{1}{6}$ 6   12   18   24   ㉚
$\frac{7}{10}$ 10   20   ㉚

6. $1\frac{13}{36}$

$\frac{5}{12}$ 12   24   ㊱
$\frac{17}{18}$ 18   ㊱

7. $1\frac{1}{36}$

$\frac{4}{9}$ 9   18   27   ㊱
$\frac{7}{12}$ 12   24   ㊱

8. $1\frac{5}{24}$

$\frac{3}{8}$ 8   16   ㉔
$\frac{5}{6}$ 6   12   18   ㉔

Page 66

1. $\frac{1}{2} = \frac{2}{4}$ 2. $\frac{1}{3} = \frac{5}{15}$
   $\frac{1}{4} = \frac{1}{4}$ $\frac{2}{5} = \frac{6}{15}$

3. $\frac{6}{7} = \frac{18}{21}$ 4. $\frac{1}{6} = \frac{4}{24}$
   $\frac{2}{21} = \frac{2}{21}$ $\frac{5}{8} = \frac{15}{24}$

5. $\frac{5}{9} = \frac{50}{90}$ 6. $\frac{1}{2}$
   $\frac{3}{10} = \frac{27}{90}$

7. $\frac{5}{8}$ 8. $\frac{8}{9}$

9. $1\frac{3}{16}$ 10. $1\frac{5}{12}$

11. $1\frac{13}{30}$ 12. $\frac{17}{20}$

13. $1\frac{5}{18}$ 14. $1\frac{1}{2}$

15. $1\frac{1}{8}$ 16. $\frac{15}{28}$

17. $\frac{8}{15}$ 18. $1\frac{5}{12}$

19. $1\frac{3}{5}$ 20. $1\frac{5}{16}$

21. $1\frac{7}{18}$ 22. $1\frac{5}{6}$

23. $1\frac{1}{6}$ 24. $\frac{20}{21}$

25. $2\frac{1}{12}$ 26. $1\frac{7}{18}$

Page 67

1. $5\frac{3}{8}$

2. $7\frac{7}{9}$

Raise $\frac{2}{3}$ to higher terms with 9 as the denominator. Add the fractions. Add the whole numbers.

$$3\frac{2}{3} = 3\frac{6}{9}$$
$$+4\frac{1}{9} = 4\frac{1}{9}$$
$$7\frac{7}{9}$$

3. $12\frac{3}{14}$ 4. $20\frac{1}{3}$

5. $7\frac{11}{12}$ 6. $5\frac{19}{24}$

7. $17\frac{31}{35}$ 8. $14\frac{19}{30}$

9. $14\frac{11}{18}$

10. $11\frac{1}{12}$

Find the LCD for the fractions. List multiples of each denominator. The smallest number on each list is 12, the LCD. Raise each fraction to higher terms with 12 as the LCD. Add the fractions. Add the whole numbers. Change the improper fraction to a mixed number.

$\frac{1}{2}$ 2 4 6 8 10 ⑫

$\frac{3}{4}$ 4 8 ⑫

$\frac{5}{6}$ 6 ⑫

$1\frac{1}{2} = 1\frac{6}{12}$

$6\frac{3}{4} = 6\frac{9}{12}$

$+2\frac{5}{6} = 2\frac{10}{12}$

$9\frac{25}{12} = 9 + 2\frac{1}{12} = 11\frac{1}{12}$

11. $19\frac{2}{5}$

Page 68

1. $10\frac{5}{24}$

2. $11\frac{1}{2}$

Line up the whole numbers. Bring down the fraction. Add the whole numbers.

$4\frac{1}{2}$
$+7$
$\overline{11\frac{1}{2}}$

3. $9\frac{7}{12}$ **4.** $9\frac{31}{56}$

5. $24\frac{1}{10}$ **6.** $6\frac{7}{12}$

7. $6\frac{1}{3}$ **8.** $24\frac{3}{4}$

9. $17\frac{13}{30}$ **10.** $9\frac{1}{2}$

11. $54\frac{19}{30}$

Page 69

1. Jorge worked 25 hours.

$3\frac{3}{4} = 3\frac{9}{12}$
$5\frac{1}{2} = 5\frac{6}{12}$
$6\frac{1}{3} = 6\frac{4}{12}$
$3\frac{3}{4} = 3\frac{9}{12}$
$+5\frac{2}{3} = 5\frac{8}{12}$
$\overline{22\frac{36}{12} = 25 \text{ hours}}$

2. Sue worked 30 hours.

$6\frac{1}{3} = 6\frac{2}{6}$
$7\phantom{\frac{1}{3}} = 7$
$6\frac{1}{2} = 6\frac{3}{6}$
$4\frac{1}{2} = 4\frac{3}{6}$
$+5\frac{2}{3} = 5\frac{4}{6}$
$\overline{28\frac{12}{6} = 30 \text{ hours}}$

3. Rachel worked 29 hours.

$6\frac{2}{3} = 6\frac{4}{6}$
$5\frac{2}{3} = 5\frac{4}{6}$
$5\frac{1}{2} = 5\frac{3}{6}$
$5\frac{1}{2} = 5\frac{3}{6}$
$+5\frac{2}{3} = 5\frac{4}{6}$
$\overline{26\frac{18}{6} = 29 \text{ hours}}$

4. 100 hours grand total

16
25
30
$+\ 29$
$\overline{100 \text{ hours}}$

Aide	November 5					Hours Worked
	M	Tu	W	Th	F	Total
Tim	$2\frac{2}{3}$	$2\frac{1}{3}$	3	$4\frac{3}{4}$	$3\frac{1}{4}$	16
Jorge	$3\frac{3}{4}$	$5\frac{1}{2}$	$6\frac{1}{3}$	$3\frac{3}{4}$	$5\frac{2}{3}$	25
Sue	$6\frac{1}{3}$	7	$6\frac{1}{2}$	$4\frac{1}{2}$	$5\frac{2}{3}$	30
Rachel	$6\frac{2}{3}$	$5\frac{2}{3}$	$5\frac{1}{2}$	$5\frac{1}{2}$	$5\frac{2}{3}$	29

Page 70

5. $\frac{3}{10}$

Set up a fraction with the number of hours June worked, 15, as the numerator. The grand total, 50, is the denominator. Reduce the fraction to lowest terms.

$\frac{15}{50} = \frac{3}{10}$

6. $\frac{1}{3}$ **7.** $\frac{1}{5}$

8. $\frac{1}{12}$

Unit 2 Review, page 71

1. $\frac{2}{3}$
2. $\frac{3}{4}$
3. $\frac{5}{8}$
4. $\frac{14}{15}$
5. $\frac{13}{16}$
6. $\frac{9}{10}$
7. $\frac{17}{25}$
8. $\frac{29}{50}$
9. $\frac{41}{100}$
10. $\frac{1}{2}$
11. $1\frac{1}{9}$
12. $1\frac{1}{11}$
13. $2\frac{1}{5}$
14. 1
15. $1\frac{1}{2}$
16. $1\frac{1}{6}$
17. $\frac{3}{5}$
18. 1
19. 6
20. $27\frac{1}{4}$
21. $50\frac{2}{3}$
22. $4\frac{3}{5}$
23. $11\frac{1}{2}$
24. $23\frac{1}{2}$
25. $24\frac{4}{5}$
26. $22\frac{1}{9}$
27. $11\frac{3}{5}$
28. $\frac{7}{10}$
29. $\frac{11}{16}$
30. $\frac{11}{18}$
31. $\frac{7}{8}$
32. $\frac{5}{14}$
33. $\frac{11}{12}$
34. $\frac{3}{10}$
35. $\frac{14}{15}$
36. $\frac{24}{25}$

Page 72

37. $\frac{1}{2}$
38. 1
39. $1\frac{1}{12}$
40. 2
41. $\frac{5}{8}$
42. $1\frac{1}{12}$
43. 1
44. $1\frac{1}{6}$
45. $1\frac{2}{5}$
46. $\frac{7}{12}$
47. $1\frac{7}{24}$
48. $\frac{25}{28}$
49. $1\frac{7}{15}$
50. $1\frac{1}{6}$
51. $1\frac{7}{12}$
52. $\frac{11}{15}$
53. $1\frac{1}{18}$
54. $1\frac{5}{24}$
55. $10\frac{1}{6}$
56. $14\frac{17}{20}$
57. $17\frac{11}{24}$
58. $11\frac{7}{12}$
59. $11\frac{3}{20}$
60. $15\frac{11}{15}$
61. $17\frac{11}{18}$
62. $46\frac{23}{24}$

Unit 3

Page 73

1. 8
2. 53
   92
   $\underline{-39}$
   53
3. 22
4. 674
5. 208
6. 575
7. 90
8. $2{,}178$
9. $1{,}759$
10. $16{,}395$

Page 74

11. $\frac{6}{8}$
12. $\frac{5}{10}$
    $\frac{1\times 5}{2\times 5} = \frac{5}{10}$
13. $\frac{8}{12}$
14. $\frac{9}{15}$
15. $\frac{9}{21}$
16. $\frac{12}{27}$
17. $\frac{3}{30}$
18. $\frac{25}{40}$
19. $\frac{28}{48}$
20. $\frac{20}{60}$
21. $\frac{1}{4} = \frac{1}{4}$
    $\frac{1}{2} = \frac{2}{4}$
22. $\frac{2}{3} = \frac{8}{12}$
    $\frac{3}{4} = \frac{9}{12}$
    $\frac{2\times 4}{3\times 4} = \frac{8}{12}$
    $\frac{3\times 3}{4\times 3} = \frac{9}{12}$
23. $\frac{5}{7} = \frac{25}{35}$
    $\frac{3}{5} = \frac{21}{35}$
24. $3\frac{3}{12} = 3\frac{3}{12}$
    $1\frac{1}{4} = 1\frac{3}{12}$
25. $2\frac{1}{10} = 2\frac{1}{10}$
    $6\frac{3}{5} = 6\frac{6}{10}$
26. $\frac{3}{8} = \frac{9}{24}$
    $\frac{5}{6} = \frac{20}{24}$
27. $2\frac{4}{9} = 2\frac{8}{18}$
    $5\frac{1}{18} = 5\frac{1}{18}$
28. $\frac{3}{10} = \frac{15}{50}$
    $\frac{3}{25} = \frac{6}{50}$
29. $\frac{2}{3} = \frac{6}{9}$
    $\frac{2}{9} = \frac{2}{9}$
30. $\frac{1}{6} = \frac{2}{12}$
    $\frac{1}{4} = \frac{3}{12}$
31. $\frac{1}{2}$

32. LT
Both fractions can be divided evenly only by 1.
33. $\frac{2}{3}$
34. $2\frac{3}{5}$
35. LT
36. LT
37. $\frac{1}{2}$
38. $\frac{1}{4}$
39. $2\frac{1}{3}$
40. $\frac{2}{7}$

Page 75

1. $\frac{3}{8}$

2.–4. *Shading should be similar to this.*

2. $\frac{2}{5}$

 −

The denominators are the same. Write the denominator under the fraction bar. Subtract the numerators. Write the difference over the denominator. Shade in 2 of the 5 parts in the final figure.
3 − 1 = 2
$\frac{3}{5} - \frac{1}{5} = \frac{2}{5}$

3. $\frac{4}{9}$

4. $\frac{1}{6}$

 −

5. $\frac{2}{9}$

6. $\frac{1}{6}$

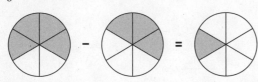

The denominators are the same. Subtract the numerators.
4 − 3 = 1
$\frac{4}{6} - \frac{3}{6} = \frac{1}{6}$

7. $\frac{3}{7}$
8. $\frac{2}{5}$

Page 76

1. $\frac{3}{14}$
2. $\frac{7}{16}$

The denominators are the same. Subtract the numerators.

$\frac{9}{16}$
$-\frac{2}{16}$
$\frac{7}{16}$

3. $\frac{6}{25}$
4. $\frac{7}{12}$
5. $\frac{2}{7}$
6. $\frac{4}{9}$
7. $\frac{7}{10}$
8. $\frac{8}{15}$
9. $\frac{7}{20}$
10. $\frac{11}{30}$
11. $\frac{3}{50}$
12. $\frac{19}{42}$
13. $\frac{3}{11}$
14. $\frac{31}{100}$
15. $\frac{23}{75}$
16. $\frac{25}{63}$
17. $\frac{9}{35}$
18. $\frac{5}{18}$
19. $\frac{1}{4}$ pound
Subtract to find how much is left.
$\frac{3}{4} - \frac{2}{4} = \frac{1}{4}$
20. $\frac{3}{8}$ yard
$\frac{7}{8} - \frac{4}{8} = \frac{3}{8}$

Page 77

1. $\frac{1}{3}$
2. $\frac{3}{4}$

The denominators are the same. Subtract the numerators. Reduce the answer to lowest terms.
$\frac{10}{12} - \frac{1}{12} = \frac{9}{12}$
$\frac{9}{12} = \frac{9 \div 3}{12 \div 3} = \frac{3}{4}$

3. $\frac{1}{4}$
4. $\frac{1}{2}$
5. $\frac{1}{2}$
6. $\frac{1}{3}$
7. $\frac{3}{7}$
8. $\frac{2}{3}$

9. $\frac{1}{2}$
10. $\frac{3}{4}$
11. $\frac{1}{5}$
12. $\frac{3}{4}$
13. $\frac{2}{3}$
14. $\frac{2}{3}$
15. $\frac{1}{6}$
16. $\frac{2}{5}$
17. $\frac{1}{2}$
18. $\frac{3}{4}$
19. $\frac{1}{2}$ pound
    Subtract to find how much candy is left.
    $\frac{3}{4} - \frac{1}{4} = \frac{2}{4} = \frac{2 \div 2}{4 \div 2} = \frac{1}{2}$
20. $\frac{1}{8}$ pound
    $\frac{9}{16} - \frac{7}{16} = \frac{2}{16} = \frac{2 \div 2}{16 \div 2} = \frac{1}{8}$

Page 78

1. $4\frac{1}{2}$
2. $5\frac{1}{4}$
   Line up the fractions and whole numbers in columns. Subtract the fractions. Subtract the whole numbers. Reduce the fraction to lowest terms.

   $10\frac{7}{8}$
   $-5\frac{5}{8}$
   $5\frac{2}{8} = 5\frac{1}{4}$

3. $6\frac{1}{2}$
4. $7\frac{1}{4}$
5. $3\frac{3}{4}$
6. $11\frac{1}{7}$
7. $7\frac{1}{2}$
8. $4\frac{1}{5}$
9. $4\frac{1}{10}$
10. $3\frac{3}{5}$
11. $11\frac{3}{16}$
12. $6\frac{1}{5}$
13. $14\frac{2}{3}$
14. $1\frac{1}{2}$ yards
    Subtract to find how many yards are left.

    $2\frac{7}{8}$
    $-1\frac{3}{8}$
    $1\frac{4}{8} = 1\frac{1}{2}$

15. $6\frac{1}{2}$ yards

    $12\frac{3}{4}$
    $-6\frac{1}{4}$
    $6\frac{2}{4} = 6\frac{1}{2}$

Page 79

1. $2\frac{1}{2}$ pounds
   Subtract how much puppy food she has from how much she needs.

   $12\frac{7}{8}$
   $-10\frac{3}{8}$
   $2\frac{4}{8} = 2\frac{1}{2}$

2. $4\frac{1}{4}$ pounds
   Subtract to find how much is left.

   $14\frac{9}{16}$
   $-10\frac{5}{16}$
   $4\frac{4}{16} = 4\frac{1}{4}$

3. $5\frac{1}{3}$ cups

   $15\frac{2}{3}$
   $-10\frac{1}{3}$
   $5\frac{1}{3}$

4. $22\frac{1}{2}$ pounds

   $50\frac{3}{4}$
   $-28\frac{1}{4}$
   $22\frac{2}{4} = 22\frac{1}{2}$

Page 80

1. $3\frac{1}{8}$
2. $9\frac{1}{2}$
   Line up the fractions. Subtract the fractions. Bring down the whole number. Reduce the fraction to lowest terms.

   $9\frac{7}{12}$
   $-\frac{1}{12}$
   $9\frac{6}{12} = 9\frac{1}{2}$

3. $11\frac{1}{3}$
4. $4\frac{1}{3}$
5. $36\frac{3}{5}$
6. $42\frac{1}{3}$
7. $50\frac{5}{16}$
8. $61\frac{2}{7}$
9. $79\frac{9}{100}$
10. $12\frac{3}{5}$

11. $20\frac{1}{6}$ 12. $14\frac{1}{2}$
13. $22\frac{7}{17}$ 14. $59\frac{2}{15}$
15. $32\frac{1}{4}$ 16. $44\frac{2}{3}$
17. $95\frac{1}{2}$ 18. $102\frac{3}{5}$

Page 81

1. $3\frac{5}{11}$
2. $8\frac{2}{3}$
   Line up the whole numbers. Bring down the fraction. Subtract the whole numbers.
   $$15\frac{2}{3}$$
   $$-\ 7$$
   $$8\frac{2}{3}$$
3. $6\frac{1}{7}$ 4. $2\frac{2}{9}$
5. $25\frac{9}{16}$ 6. $65\frac{22}{35}$
7. $19\frac{9}{10}$ 8. $23\frac{27}{50}$
9. $79\frac{73}{100}$ 10. $9\frac{7}{13}$
11. $15\frac{10}{21}$ 12. $7\frac{9}{20}$
13. $4\frac{12}{25}$
14. $1\frac{1}{2}$ pounds
    Subtract to find how much he has left.
    $$3\frac{1}{2}$$
    $$-\ 2$$
    $$1\frac{1}{2}$$
15. $\frac{3}{4}$ gallon
    $$1\frac{3}{4}$$
    $$-\ 1$$
    $$\frac{3}{4}$$

Page 82

1. $\frac{3}{4}$ 2. $\frac{1}{5}$
3. $\frac{2}{3}$ 4. $\frac{1}{2}$
5. $\frac{8}{9}$ 6. $\frac{4}{5}$
7. $3\frac{1}{3}$ 8. $\frac{1}{2}$
9. $6\frac{3}{4}$ 10. $\frac{4}{7}$
11. $6\frac{3}{5}$ 12. $4\frac{1}{2}$
13. $\frac{1}{3}$ 14. $32\frac{1}{4}$
15. $1\frac{2}{3}$ 16. $\frac{1}{5}$
17. $10\frac{10}{13}$ 18. $\frac{2}{7}$
19. $\frac{27}{35}$ 20. $\frac{1}{9}$
21. $11\frac{2}{3}$ 22. $9\frac{1}{7}$
23. $\frac{1}{7}$ 24. $5\frac{9}{10}$
25. $\frac{1}{4}$ 26. $13\frac{1}{2}$
27. $\frac{3}{7}$ 28. $16\frac{3}{5}$
29. $4\frac{41}{49}$ 30. $\frac{2}{3}$
31. $8\frac{1}{8}$ 32. $50\frac{99}{100}$
33. $8\frac{2}{3}$ 34. $\frac{9}{20}$
35. $\frac{1}{5}$ 36. $3\frac{1}{3}$

Page 83

1. $4\frac{1}{4}$ inches
   Point B is $\frac{5}{8}$ inch from 0. Point I is $4\frac{7}{8}$ inches from 0. Subtract. Reduce to lowest terms.
   $$4\frac{7}{8}$$
   $$-\ \frac{5}{8}$$
   $$4\frac{2}{8} = 4\frac{1}{4}$$
2. $5\frac{1}{2}$ inches
   $$5\frac{3}{4}$$
   $$-\ \frac{1}{4}$$
   $$5\frac{2}{4} = 5\frac{1}{2}$$
3. $3\frac{1}{4}$ inches   4. $\frac{1}{2}$ inch
   $$3\frac{7}{8}$$ $$1\frac{1}{2}$$
   $$-\ \frac{5}{8}$$ $$-\ 1$$
   $$3\frac{2}{8} = 3\frac{1}{4}$$ $$\frac{1}{2}$$

Page 84

5. 1 inch   6. 3 inches
   $$2\frac{1}{2}$$ $$5\frac{3}{4}$$
   $$-\ 1\frac{1}{2}$$ $$-\ 2\frac{3}{4}$$
   $$1$$ $$3$$

7. $1\frac{3}{4}$ inches 8. 1 inch

$$2\frac{3}{4}$$
$$-1$$
$$\overline{1\frac{3}{4}}$$

$$4\frac{7}{8}$$
$$-3\frac{7}{8}$$
$$\overline{1}$$

9. $1\frac{1}{2}$ inches 10. 4 inches

$$5\frac{3}{4}$$
$$-4\frac{1}{4}$$
$$\overline{1\frac{2}{4}} = 1\frac{1}{2}$$

$$4\frac{1}{4}$$
$$-\frac{1}{4}$$
$$\overline{4}$$

11. 1 inch
Subtract to find how much molding is left.

$$4\frac{1}{2}$$
$$-3\frac{1}{2}$$
$$\overline{1}$$

12. $3\frac{1}{2}$ inches

$$9\frac{3}{4}$$
$$-6\frac{1}{4}$$
$$\overline{3\frac{2}{4}} = 3\frac{1}{2}$$

13. $2\frac{1}{2}$ inches
Subtract to find how much Helen's son grew.

$$36\frac{3}{4}$$
$$-34\frac{1}{4}$$
$$\overline{2\frac{2}{4}} = 2\frac{1}{2}$$

14. $3\frac{1}{2}$ inches

$$52\frac{5}{8}$$
$$-49\frac{1}{8}$$
$$\overline{3\frac{4}{8}} = 3\frac{1}{2}$$

Page 85

1. $\frac{3}{3}$
2. $\frac{5}{5}$
Change 1 to an improper fraction with 5 as the denominator and the numerator.
3. $\frac{10}{10}$ 4. $\frac{7}{7}$
5. $1\frac{2}{2}$
6. $2\frac{5}{5}$
Borrow 1 from 3. 3 − 1 = 2. Change 1 to an improper fraction with 5 as the denominator. Write $\frac{5}{5}$ next to the 2.
7. $6\frac{10}{10}$ 8. $8\frac{12}{12}$
9. $5\frac{7}{7}$ 10. $3\frac{6}{6}$
11. $4\frac{4}{4}$ 12. $7\frac{9}{9}$
13. $11\frac{3}{3}$ 14. $14\frac{16}{16}$
15. $9\frac{14}{14}$ 16. $16\frac{15}{15}$
17. $26\frac{5}{5}$
18. $98\frac{12}{12}$
Borrow 1 from 99. 99 − 1 = 98. Change 1 to an improper fraction with 12 as the denominator. Write $\frac{12}{12}$ next to the 98.
19. $13\frac{17}{17}$ 20. $18\frac{20}{20}$
21. $120\frac{8}{8}$ 22. $135\frac{11}{11}$
23. $139\frac{18}{18}$ 24. $58\frac{21}{21}$
25. $99\frac{25}{25}$ 26. $89\frac{30}{30}$
27. $100\frac{49}{49}$ 28. $127\frac{100}{100}$

Page 86

1. $\frac{9}{9}$
2. $\frac{6}{6}$
Change 1 to an improper fraction with 6 as the denominator and the numerator.
3. $\frac{10}{10}$ 4. $\frac{15}{15}$
5. $\frac{50}{50}$ 6. $\frac{4}{7}$
7. $\frac{5}{12}$
Change 1 to an improper fraction with 12 as the denominator. Subtract the fractions.

$$\frac{12}{12} - \frac{7}{12} = \frac{5}{12}$$

8. $\frac{4}{15}$ 9. $\frac{13}{25}$
10. $\frac{1}{6}$ 11. $\frac{9}{19}$
12. $\frac{5}{11}$ 13. $\frac{3}{17}$

186

14. $\frac{1}{30}$ 15. $\frac{1}{3}$
16. $\frac{1}{10}$ 17. $\frac{5}{8}$
18. $\frac{13}{20}$

11. $\frac{20}{21}$ 12. $\frac{51}{76}$
13. $\frac{89}{100}$ 14. $\frac{6}{7}$
15. $\frac{4}{5}$ 16. $\frac{3}{4}$
17. $\frac{3}{5}$ 18. $\frac{37}{54}$

Page 87

1. $1\frac{1}{3}$
2. $4\frac{1}{4}$
   Borrow 1 from 5. $5 - 1 = 4$. Change 1 to a fraction with 4 as the denominator. Subtract the fractions. Bring down the whole number.

   $$\begin{aligned} 5 &= 4\frac{4}{4} \\ -\frac{3}{4} &= \frac{3}{4} \\ \hline &4\frac{1}{4} \end{aligned}$$

3. $3\frac{7}{10}$ 4. $7\frac{5}{9}$
5. $29\frac{7}{18}$ 6. $25\frac{1}{20}$
7. $36\frac{8}{21}$ 8. $39\frac{22}{25}$
9. $52\frac{23}{35}$ 10. $9\frac{2}{5}$
11. $12\frac{1}{6}$ 12. $14\frac{5}{12}$
13. $20\frac{13}{15}$ 14. $93\frac{39}{49}$
15. $29\frac{11}{29}$ 16. $75\frac{47}{50}$
17. $80\frac{38}{75}$ 18. $24\frac{19}{100}$

Page 88

1. $\frac{2}{3}$
2. $\frac{4}{7}$
   Change 1 to an improper fraction with 7 as the denominator. Add $\frac{7}{7} + \frac{2}{7} = \frac{9}{7}$. Subtract.

   $$\begin{aligned} 1\frac{2}{7} &= \frac{9}{7} \\ -\frac{5}{7} &= \frac{5}{7} \\ \hline &\frac{4}{7} \end{aligned}$$

3. $\frac{1}{2}$ 4. $\frac{8}{11}$
5. $\frac{14}{15}$ 6. $\frac{11}{21}$
7. $\frac{7}{8}$ 8. $\frac{17}{25}$
9. $\frac{16}{21}$ 10. $\frac{2}{3}$

Page 89

1. $10\frac{10}{11}$
2. $30\frac{5}{8}$
   Since $\frac{5}{8}$ is greater than $\frac{2}{8}$, borrow 1 from 31. Change the borrowed 1 to an improper fraction with 8 as the denominator. $1 = \frac{8}{8}$. Add the mixed number and the fraction. $30\frac{2}{8} + \frac{8}{8} = 30\frac{10}{8}$. Subtract the fractions. Bring down the whole number. Reduce.

   $$\begin{aligned} 31\frac{2}{8} &= 30\frac{10}{8} \\ -\frac{5}{8} &= \frac{5}{8} \\ \hline & 30\frac{5}{8} \end{aligned}$$

3. $11\frac{1}{3}$ 4. $19\frac{3}{4}$
5. $44\frac{7}{15}$ 6. $6\frac{7}{9}$
7. $3\frac{3}{5}$
   Since $\frac{7}{10}$ is greater than $\frac{3}{10}$, borrow 1 from 4. Change the borrowed 1 to an improper fraction with 10 as the denominator. $1 = \frac{10}{10}$. Add the mixed number and the fraction. $3\frac{3}{10} + \frac{10}{10} = 3\frac{13}{10}$. Subtract the fractions. Bring down the whole number. Reduce to lowest terms.

   $$\begin{aligned} 4\frac{3}{10} &= 3\frac{13}{10} \\ -\frac{7}{10} &= \frac{7}{10} \\ \hline & 3\frac{6}{10} = 3\frac{3}{5} \end{aligned}$$

8. $5\frac{1}{2}$ 9. $1\frac{4}{7}$
10. $16\frac{20}{21}$ 11. $39\frac{7}{9}$
12. $12\frac{13}{15}$ 13. $71\frac{47}{50}$
14. $17\frac{61}{63}$ 15. $24\frac{7}{8}$
16. $30\frac{2}{5}$ 17. $5\frac{5}{9}$
18. $11\frac{1}{5}$

Page 90

1. $3\frac{2}{3}$
2. $1\frac{1}{2}$
   Since $\frac{3}{4}$ is greater than $\frac{1}{4}$, borrow 1 from the 5. Change the borrowed 1 to an improper fraction with 4 as the denominator. Add the mixed number and the fraction. $4\frac{1}{4} + \frac{4}{4} = 4\frac{5}{4}$.
   Subtract the fractions. Subtract the whole numbers. Reduce.

   $$5\frac{1}{4} = 4\frac{5}{4}$$
   $$-3\frac{3}{4} = 3\frac{3}{4}$$
   $$1\frac{2}{4} = 1\frac{1}{2}$$

3. $4\frac{6}{7}$
4. $1\frac{4}{5}$
5. $4\frac{1}{2}$
6. $\frac{2}{3}$
7. $9\frac{9}{10}$
8. $9\frac{4}{5}$
9. $5\frac{15}{19}$
10. $1\frac{1}{2}$ gallons
    Subtract to find how many gallons are left in the gas tank.

    $$18\frac{1}{4} = 17\frac{5}{4}$$
    $$-16\frac{3}{4} = 16\frac{3}{4}$$
    $$1\frac{2}{4} = 1\frac{1}{2}$$

11. $6\frac{3}{5}$ inches
    Subtract to find how much higher her record was this year than last year.

    $$59\frac{3}{10} = 58\frac{13}{10}$$
    $$-52\frac{7}{10} = 52\frac{7}{10}$$
    $$6\frac{6}{10} = 6\frac{3}{5}$$

Page 91

1. 13 inches
   Subtract the length of the second support, $2\frac{1}{2}$ inches, from the amount left after you cut off the first support.

   $$15\frac{1}{2}$$
   $$-2\frac{1}{2}$$
   $$13$$

2. $5\frac{1}{2}$ inches
   Subtract to find how much wood is left.

   $$13 = 12\frac{2}{2}$$
   $$-7\frac{1}{2} = 7\frac{1}{2}$$
   $$5\frac{1}{2}$$

3. $\frac{1}{4}$ inch
   Subtract the length the perch should be, $2\frac{3}{4}$ inches, from the length of the rod for the perch, 3 inches.

   $$3 = 2\frac{4}{4}$$
   $$-2\frac{3}{4} = 2\frac{3}{4}$$
   $$\frac{1}{4}$$

4. $\frac{1}{2}$ inch
   Subtract the length of the bottom of the feeder, $7\frac{1}{2}$ inches, from the length of the tin can, 8 inches.

   $$8 = 7\frac{2}{2}$$
   $$-7\frac{1}{2} = 7\frac{1}{2}$$
   $$\frac{1}{2}$$

Page 92

1. 1
2. 7
3. $1\frac{1}{2}$
4. $1\frac{1}{3}$
5. $4\frac{1}{2}$
6. 13
7. $12\frac{3}{7}$
8. 1
9. $\frac{3}{4}$
10. $8\frac{2}{2}$
11. $5\frac{3}{3}$
12. $17\frac{9}{9}$
13. $19\frac{10}{10}$
14. $28\frac{35}{35}$
15. $\frac{1}{4}$
16. $1\frac{3}{5}$
17. $14\frac{17}{20}$
18. $19\frac{13}{16}$
19. $4\frac{3}{5}$
20. 21
21. $4\frac{10}{11}$
22. $2\frac{4}{5}$

23. $4\frac{4}{9}$
24. $30\frac{1}{2}$
25. $\frac{3}{4}$
26. 8
27. $18\frac{19}{21}$
28. $\frac{2}{3}$
29. $\frac{4}{5}$
30. $31\frac{3}{8}$
31. $1\frac{1}{8}$
32. $27\frac{3}{50}$

Page 93

1. $\frac{7}{10}$
2. $\frac{5}{12}$

The denominators are different. Since 4 divides evenly into 12, write 12 as the common denominator. Raise $\frac{1}{4}$ to higher terms with 12 as the denominator. Subtract the numerators. Write the difference over the denominator.

$$\begin{aligned}\frac{8}{12} &= \frac{8}{12}\\ -\frac{1}{4} &= \frac{3}{12}\\ \hline &\frac{5}{12}\end{aligned}$$

3. $\frac{4}{9}$
4. $\frac{9}{14}$
5. $\frac{4}{15}$
6. $\frac{5}{18}$
7. $\frac{11}{20}$
8. $\frac{3}{16}$
9. $\frac{3}{20}$
10. $\frac{1}{9}$
11. $\frac{1}{5}$
12. $\frac{1}{5}$
13. $\frac{2}{7}$

Page 94

1. $\frac{1}{4}$
2. $\frac{3}{8}$

Raise $\frac{1}{2}$ to higher terms with 8 as the denominator. Subtract. Reduce.

$$\begin{aligned}\frac{1}{2} &= \frac{4}{8}\\ -\frac{1}{8} &= \frac{1}{8}\\ \hline &\frac{3}{8}\end{aligned}$$

3. $\frac{5}{9}$
4. $\frac{1}{2}$
5. $\frac{9}{16}$
6. $\frac{3}{4}$
7. $\frac{5}{18}$
8. $\frac{1}{3}$
9. $\frac{5}{8}$
10. $\frac{13}{25}$
11. $\frac{2}{3}$
12. $\frac{2}{15}$
13. $\frac{9}{20}$
14. $\frac{1}{4}$ teaspoon

Subtract to find how much more medicine Janie gave her son.

$$\begin{aligned}\frac{1}{2} &= \frac{2}{4}\\ -\frac{1}{4} &= \frac{1}{4}\\ \hline &\frac{1}{4}\end{aligned}$$

15. $\frac{1}{10}$ mile

Subtract to find how much farther Viola swam.

$$\begin{aligned}\frac{4}{5} &= \frac{8}{10}\\ -\frac{7}{10} &= \frac{7}{10}\\ \hline &\frac{1}{10}\end{aligned}$$

Page 95

1. $\frac{1}{10}$
2. $\frac{5}{12}$

Multiply to find the common denominator. $3 \times 4 = 12$. Raise each fraction to higher terms with 12 as the denominator. Subtract.

$$\begin{aligned}\frac{3}{4} &= \frac{9}{12}\\ -\frac{1}{3} &= \frac{4}{12}\\ \hline &\frac{5}{12}\end{aligned}$$

3. $\frac{1}{6}$
4. $\frac{2}{15}$
5. $\frac{19}{30}$
6. $\frac{1}{24}$
7. $\frac{1}{14}$
8. $\frac{11}{30}$
9. $\frac{7}{20}$
10. $\frac{11}{42}$
11. $\frac{8}{21}$
12. $\frac{3}{14}$
13. $\frac{5}{36}$
14. $\frac{6}{35}$
15. $\frac{5}{18}$
16. $\frac{11}{20}$
17. $\frac{7}{20}$
18. $\frac{7}{24}$

Page 96

1. $\frac{7}{12}$
2. $\frac{7}{18}$

   List the multiples of each denominator. The smallest number on both lists is 18. Raise each fraction to higher terms with 18 as the LCD. Subtract.

   $\frac{5}{6}$ 12   ⑱   24   30

   $\frac{4}{9}$ ⑱   27   36

   $\frac{5}{6} = \frac{15}{18}$
   $-\frac{4}{9} = \frac{8}{18}$
   $\phantom{-\frac{4}{9} = }\frac{7}{18}$

3. $\frac{9}{20}$
4. $\frac{7}{24}$
5. $\frac{11}{36}$
6. $\frac{7}{30}$
7. $\frac{7}{36}$
8. $\frac{17}{45}$
9. $\frac{35}{48}$

Page 97

1. $\frac{17}{40}$
2. $\frac{17}{24}$
3. $\frac{1}{8}$
4. $\frac{9}{28}$
5. $\frac{11}{40}$
6. $\frac{6}{35}$
7. $\frac{1}{10}$
8. $\frac{13}{42}$
9. $\frac{5}{14}$
10. $\frac{17}{39}$
11. $\frac{7}{24}$
12. $\frac{5}{12}$
13. $\frac{5}{9}$
14. $\frac{4}{9}$
15. $\frac{9}{20}$
16. $\frac{7}{12}$
17. $\frac{1}{18}$
18. $\frac{7}{30}$
19. $\frac{17}{45}$
20. $\frac{11}{36}$
21. $\frac{1}{20}$
22. $\frac{1}{5}$
23. $\frac{1}{12}$
24. $\frac{17}{30}$
25. $\frac{17}{36}$
26. $\frac{2}{3}$

Page 98

1. $\frac{2}{5}$ mile

   Subtract to find how much farther Junction is from Fort Smith.

   $3\frac{7}{10}$
   $-3\frac{3}{10}$
   $\frac{4}{10} = \frac{2}{5}$

2. $8\frac{7}{10}$ miles

   Subtract the distance from Fort Smith to Newton from the distance from Fort Smith to Keene.

   $12 \phantom{\frac{3}{10}} = 11\frac{10}{10}$
   $-3\frac{3}{10} = 3\frac{3}{10}$
   $\phantom{-12\frac{3}{10} = }8\frac{7}{10}$

Page 99

3. $1\frac{1}{10}$ miles

   Subtract to find how many more miles the distance is from Fort Smith to Keene.

   $12 \phantom{\frac{9}{10}} = 11\frac{10}{10}$
   $-10\frac{9}{10} = 10\frac{9}{10}$
   $\phantom{-10\frac{9}{10} = }1\frac{1}{10}$

4. $7\frac{3}{10}$ miles

   Add the distance from Fort Smith to Junction to the distance from Junction to Newton.

   $4 \phantom{\frac{7}{10}} = 3\frac{10}{10}$
   $+3\frac{7}{10} = 3\frac{7}{10}$
   $\phantom{+3\frac{7}{10} = }6\frac{17}{10} = 7\frac{7}{10}$

5. $7\frac{9}{10}$ miles

   Subtract to find how many miles he will save.

   $15 \phantom{\frac{1}{10}} = 14\frac{10}{10}$
   $-7\frac{1}{10} = 7\frac{1}{10}$
   $\phantom{-15\frac{1}{10} = }7\frac{9}{10}$

6. $4\frac{1}{2}$ miles

   Subtract to find how much shorter it is to cross the lake by boat.

   $$\begin{array}{r} 10\phantom{\tfrac{2}{2}} = 9\tfrac{2}{2} \\ -\ 5\tfrac{1}{2} = 5\tfrac{1}{2} \\ \hline 4\tfrac{1}{2} \end{array}$$

7. $\frac{9}{10}$ mile

   Subtract the distance from Wood to Keene from the distance from Wood to Newton.

   $$\begin{array}{r} 6\phantom{\tfrac{10}{10}} = 5\tfrac{10}{10} \\ -5\tfrac{1}{10} = 5\tfrac{1}{10} \\ \hline \tfrac{9}{10} \end{array}$$

8. She lives closer to Keene.

   Subtract the distance Bob's sister lives from Keene from the total distance between Wood and Keene. She lives 3 miles from Wood.

   $$\begin{array}{r} 5\tfrac{1}{10} \\ -\ 2\tfrac{1}{10} \\ \hline 3 \end{array}$$

9. $2\frac{4}{5}$ miles

   Subtract the distance at which Bob's car broke down from Keene from the total distance between Keene and Wood.

   $$\begin{array}{r} 5\tfrac{1}{10} = 4\tfrac{11}{10} \\ -2\tfrac{3}{10} = 2\tfrac{3}{10} \\ \hline 2\tfrac{8}{10} = 2\tfrac{4}{5} \end{array}$$

10. $38\frac{2}{5}$ miles

    Subtract to find how many miles he drove this week.

    $$\begin{array}{r} 42\tfrac{1}{10} = 41\tfrac{11}{10} \\ -\ 3\tfrac{7}{10} = \ 3\tfrac{7}{10} \\ \hline 38\tfrac{4}{10} = 38\tfrac{2}{5} \end{array}$$

Page 100

1. $2\frac{1}{8}$

2. $4\frac{1}{2}$

   Since 3 divides evenly into 6, use 6 as the common denominator. Raise $\frac{2}{3}$ to higher terms with 6 as the denominator. Subtract the fractions. Subtract the whole numbers. Reduce to lowest terms.

   $$\begin{array}{r} 7\tfrac{2}{3} = 7\tfrac{4}{6} \\ -\ 3\tfrac{1}{6} = 3\tfrac{1}{6} \\ \hline 4\tfrac{3}{6} = 4\tfrac{1}{2} \end{array}$$

3. $3\frac{1}{4}$ 4. $2\frac{1}{2}$

5. $10\frac{5}{12}$ 6. $5\frac{7}{30}$

7. $12\frac{11}{30}$ 8. $12\frac{1}{18}$

9. $7\frac{1}{8}$

10. $2\frac{1}{12}$ yards

    Subtract the amount of material she used from the total amount she bought.

    $$\begin{array}{r} 3\tfrac{3}{4} = 3\tfrac{9}{12} \\ -\ 1\tfrac{2}{3} = 1\tfrac{8}{12} \\ \hline 2\tfrac{1}{12} \end{array}$$

11. $3\frac{5}{24}$ yards

    $$\begin{array}{r} 4\tfrac{7}{8} = 4\tfrac{21}{24} \\ -1\tfrac{2}{3} = 1\tfrac{16}{24} \\ \hline 3\tfrac{5}{24} \end{array}$$

Page 101

1. $\frac{7}{9}$

2. $2\frac{5}{8}$

   Since 2 will divide evenly into 8, use 8 as the common denominator. Raise $\frac{1}{2}$ to higher terms with 8 as the denominator. Since $\frac{7}{8}$ is greater than $\frac{4}{8}$, borrow 1 from the 9. Change it to $\frac{8}{8}$. Subtract.

   $$\begin{array}{r} 9\tfrac{1}{2} = 9\tfrac{4}{8} = 8\tfrac{12}{8} \\ -\ 6\tfrac{7}{8} = 6\tfrac{7}{8} = 6\tfrac{7}{8} \\ \hline 2\tfrac{5}{8} \end{array}$$

3. $7\frac{19}{24}$ 4. $8\frac{17}{36}$

5. $14\frac{1}{2}$
6. $7\frac{11}{24}$
7. $13\frac{15}{22}$
8. $15\frac{13}{15}$
9. $\frac{11}{20}$
10. $35\frac{14}{25}$
11. $36\frac{13}{18}$
12. $56\frac{7}{10}$

Page 102

1. $3\frac{14}{15}$
2. $2\frac{20}{21}$

   Find a common denominator.
   $7 \times 3 = 21$. Raise both fractions to higher terms with a denominator of 21. Since $\frac{7}{21}$ is greater than $\frac{6}{21}$, borrow 1 from 4.
   $4\frac{6}{21} = 3\frac{27}{21}$. Subtract the fractions. Subtract the whole numbers.

   $$4\frac{2}{7} = 4\frac{6}{21} = 3\frac{27}{21}$$
   $$-1\frac{1}{3} = 1\frac{7}{21} = 1\frac{7}{21}$$
   $$\overline{\phantom{-1\frac{1}{3} = 1\frac{7}{21} = }2\frac{20}{21}}$$

3. $6\frac{1}{10}$
4. $6\frac{1}{4}$
5. $\frac{13}{24}$
6. $3\frac{7}{10}$
7. $12\frac{9}{10}$
8. $5\frac{20}{21}$
9. $5\frac{14}{15}$

Page 103

1. $\frac{4}{15}$
2. $2\frac{21}{40}$

   Find a common denominator.
   $5 \times 8 = 40$. Since $\frac{35}{40}$ is greater than $\frac{16}{40}$, borrow 1 from the 3. Add the mixed number and the fraction.
   $2\frac{16}{40} + \frac{40}{40} = 2\frac{56}{40}$. Subtract the fractions. Bring down the whole number.

   $$3\frac{2}{5} = 3\frac{16}{40} = 2\frac{56}{40}$$
   $$-\frac{7}{8} = \frac{35}{40} = \frac{35}{40}$$
   $$\overline{\phantom{-\frac{7}{8} = \frac{35}{40} = }2\frac{21}{40}}$$

3. $\frac{19}{28}$
4. $3\frac{5}{6}$
5. $8\frac{7}{8}$
6. $\frac{17}{18}$
7. $4\frac{1}{2}$
8. $9\frac{7}{12}$

9. $1\frac{31}{33}$

Unit 3 Review, page 104

1. $\frac{9}{13}$
2. $\frac{3}{5}$
3. $\frac{1}{4}$
4. $\frac{1}{5}$
5. $\frac{1}{3}$
6. $\frac{1}{3}$
7. $\frac{2}{3}$
8. $\frac{1}{3}$
9. $\frac{1}{2}$
10. $8\frac{1}{5}$
11. $2\frac{1}{2}$
12. $14\frac{3}{8}$
13. $3\frac{2}{3}$
14. $6\frac{1}{5}$
15. $10\frac{1}{4}$
16. $1\frac{3}{5}$
17. $18\frac{12}{35}$
18. $2\frac{3}{8}$
19. $5\frac{1}{10}$
20. $8\frac{2}{3}$
21. $30\frac{47}{50}$
22. $5\frac{1}{6}$
23. $8\frac{10}{11}$
24. $\frac{1}{6}$
25. $\frac{3}{10}$
26. $\frac{1}{4}$
27. $\frac{2}{3}$

Page 105

28. $3\frac{1}{2}$
29. $5\frac{2}{3}$
30. $2\frac{1}{4}$
31. $5\frac{7}{9}$
32. $7\frac{7}{10}$
33. $9\frac{3}{5}$
34. $17\frac{5}{6}$
35. $10\frac{3}{7}$
36. $12\frac{7}{12}$
37. $3\frac{2}{3}$
38. $\frac{2}{3}$
39. $6\frac{1}{2}$
40. $8\frac{10}{11}$
41. $\frac{7}{9}$
42. $5\frac{3}{5}$
43. $1\frac{3}{5}$
44. $\frac{2}{3}$
45. $4\frac{4}{5}$
46. $\frac{1}{2}$
47. $\frac{1}{14}$
48. $\frac{1}{6}$
49. $\frac{2}{5}$
50. $\frac{1}{6}$
51. $\frac{3}{20}$
52. $\frac{7}{30}$
53. $\frac{11}{42}$
54. $\frac{5}{24}$
55. $\frac{7}{12}$

56. $\frac{1}{20}$ **57.** $\frac{13}{24}$
58. $\frac{13}{40}$

Page 106

59. $3\frac{1}{12}$ **60.** $3\frac{1}{2}$
61. $11\frac{7}{12}$ **62.** $1\frac{1}{4}$
63. $5\frac{1}{30}$ **64.** $19\frac{1}{20}$
65. $5\frac{4}{9}$ **66.** $5\frac{7}{12}$
67. $7\frac{1}{15}$ **68.** $8\frac{13}{14}$
69. $1\frac{7}{8}$ **70.** $12\frac{7}{20}$
71. $3\frac{13}{36}$ **72.** $12\frac{26}{33}$
73. $13\frac{17}{30}$ **74.** $\frac{35}{36}$
75. $4\frac{31}{40}$ **76.** $1\frac{47}{60}$

Unit 4

Page 107

1. 12 **2.** 12
              6
           × 2
           ———
             12
3. 5 **4.** 18
5. 21 **6.** 20
7. 6 **8.** 40
9. 27 **10.** 9
11. 42 **12.** 32
13. 63 **14.** 15
15. 10 **16.** 16

Page 108

17. $\frac{1}{6}$ **18.** $\frac{3}{10}$
                       $\frac{6 \div 2}{20 \div 2} = \frac{3}{10}$
19. LT **20.** $\frac{9}{16}$
21. $\frac{1}{3}$ **22.** $\frac{3}{7}$
23. LT **24.** $\frac{4}{5}$
25. $\frac{4}{15}$ **26.** $\frac{1}{3}$
27. $\frac{4}{5}$ **28.** LT
29. $\frac{1}{7}$ **30.** $\frac{1}{2}$
31. $\frac{1}{5}$ **32.** $5\frac{1}{2}$

33. $3\frac{1}{5}$

Divide the numerator, 16, by the denominator, 5. Write the remainder as a fraction.

$$5\overline{)16} = 3\frac{1}{5}$$
$$-15$$
$$1$$

34. 18 **35.** $9\frac{2}{3}$
36. $8\frac{1}{2}$ **37.** $3\frac{1}{3}$
38. $3\frac{3}{8}$ **39.** 6
40. $5\frac{4}{9}$ **41.** $5\frac{3}{10}$
42. 6 **43.** 2
44. $6\frac{1}{4}$ **45.** $2\frac{1}{7}$
46. $1\frac{1}{20}$ **47.** $3\frac{1}{8}$
48. 4 **49.** $11\frac{1}{2}$
50. 3 **51.** $4\frac{2}{13}$

Page 109

1. $\frac{1}{6}$

2.–6. Shading should be similar to this.

2. $\frac{1}{8}$

Set up the problem. $\frac{1}{2} \times \frac{1}{4}$.
Multiply the numerators.
Multiply the denominators.

$$\frac{1}{2} \times \frac{1}{4} = \frac{1}{8}$$

3. $\frac{3}{8}$

4. $\frac{2}{9}$

5. $\frac{3}{10}$

6. $\frac{3}{12}$ or $\frac{1}{4}$ (lowest terms)

7. $\frac{3}{16}$ 8. $\frac{7}{24}$

9. $\frac{1}{10}$

Page 110

1. $\frac{1}{10}$
2. $\frac{5}{12}$
   Multiply the numerators.
   Multiply the denominators.
   $\frac{1}{2} \times \frac{5}{6} = \frac{5}{12}$
3. $\frac{1}{7}$ 4. $\frac{1}{16}$
5. $\frac{1}{2}$ 6. $\frac{4}{21}$
7. $\frac{9}{20}$ 8. $\frac{4}{81}$
9. $\frac{12}{49}$ 10. $\frac{5}{24}$
11. $\frac{4}{25}$ 12. $\frac{1}{16}$
13. $\frac{5}{27}$ 14. $\frac{1}{9}$
15. $\frac{2}{7}$ 16. $\frac{21}{32}$
17. $\frac{5}{36}$ 18. $\frac{1}{4}$
19. $\frac{30}{49}$ 20. $\frac{1}{5}$
21. $\frac{9}{16}$ 22. $\frac{1}{42}$
23. $\frac{2}{3}$ 24. $\frac{15}{64}$
25. $\frac{3}{8}$ pound   26. $\frac{1}{6}$ quart
    $\frac{3}{4} \times \frac{1}{2} = \frac{3}{8}$ $\frac{1}{2} \times \frac{1}{3} = \frac{1}{6}$

Page 111

1. $\frac{1}{10}$
   Cassettes are $\frac{2}{5}$ of the total sales. $\frac{1}{4}$ of the cassettes sold are rap. Multiply.
   $\frac{1}{4} \times \frac{2}{5} = \frac{2}{20} = \frac{2 \div 2}{20 \div 2} = \frac{1}{10}$
2. $\frac{1}{20}$
   $\frac{1}{8} \times \frac{2}{5} = \frac{2}{40} = \frac{2 \div 2}{40 \div 2} = \frac{1}{20}$
3. $\frac{3}{100}$
   $\frac{1}{5} \times \frac{3}{20} = \frac{3}{100}$
4. $\frac{9}{100}$
   $\frac{3}{5} \times \frac{3}{20} = \frac{9}{100}$
5. $\frac{3}{20}$
   $\frac{1}{3} \times \frac{9}{20} = \frac{9}{60} = \frac{9 \div 3}{60 \div 3} = \frac{3}{20}$
6. $\frac{1}{5}$
   $\frac{4}{9} \times \frac{9}{20} = \frac{36}{180} = \frac{36 \div 36}{180 \div 36} = \frac{1}{5}$

Page 112

1. $\frac{1}{18}$
2. $\frac{1}{15}$
   Multiply the first two numerators.
   $1 \times 2 = 2$. Multiply 2 by the last numerator. $2 \times 1 = 2$. Multiply the first two denominators. $2 \times 5 = 10$. Multiply 10 by the last denominator. $10 \times 3 = 30$. Reduce the answer to lowest terms.
   $\frac{1}{2} \times \frac{2}{5} \times \frac{1}{3} = \frac{2}{30} = \frac{2 \div 2}{30 \div 2} = \frac{1}{15}$
3. $\frac{3}{8}$ 4. $\frac{1}{30}$
5. $\frac{4}{21}$ 6. $\frac{5}{32}$
7. $\frac{3}{50}$ 8. $\frac{1}{27}$
9. $\frac{4}{9}$ 10. $\frac{2}{15}$
11. $\frac{1}{7}$ 12. $\frac{1}{8}$
13. $\frac{6}{49}$ 14. $\frac{7}{50}$
15. $\frac{3}{100}$ 16. $\frac{5}{16}$
17. $\frac{1}{15}$ 18. $\frac{1}{30}$
19. $\frac{1}{12}$ 20. $\frac{3}{26}$

21. $\frac{3}{32}$

Page 113

1. $\frac{1}{15}$
2. $\frac{3}{14}$

   Cancel by dividing the numerator of the second fraction, 4, and the denominator of the first fraction, 8, by 4. Multiply the numerators of the new fractions. Multiply the denominators of the new fractions.

   $$\frac{3}{8} \times \frac{4}{7} = \frac{3}{\cancel{8}_2} \times \frac{\cancel{4}^1}{7} = \frac{3}{14}$$

3. $\frac{3}{7}$
4. $\frac{4}{7}$
5. $\frac{9}{14}$
6. $\frac{6}{49}$
7. $\frac{1}{6}$
8. $\frac{8}{39}$
9. $\frac{3}{32}$
10. $\frac{2}{17}$
11. $\frac{8}{63}$
12. $\frac{27}{50}$
13. $\frac{5}{54}$
14. $\frac{3}{28}$
15. $\frac{22}{75}$
16. $\frac{4}{33}$
17. $\frac{7}{55}$
18. $\frac{9}{28}$
19. $\frac{1}{36}$
20. $\frac{8}{105}$
21. $\frac{4}{45}$
22. $\frac{3}{35}$
23. $\frac{3}{10}$
24. $\frac{14}{19}$

Page 114

1. $\frac{1}{2}$
2. $\frac{1}{2}$

   Cancel once by dividing the numerator of the first fraction, 4, and the denominator of the second fraction, 16, by 4. Write the new numbers. Cancel again by dividing the numerator of the second fraction, 10, and the denominator of the first fraction, 5, by 5. Multiply the new numerators and denominators.

   $$\frac{\cancel{4}^1}{\cancel{5}_1} \times \frac{\cancel{10}^2}{\cancel{16}_4} = \frac{2}{4} = \frac{1}{2}$$

3. $\frac{3}{5}$
4. $\frac{3}{7}$
5. $\frac{7}{8}$
6. $\frac{2}{9}$
7. $\frac{2}{5}$
8. $\frac{3}{4}$
9. $\frac{2}{21}$
10. $\frac{2}{3}$
11. $\frac{1}{3}$
12. $\frac{4}{15}$
13. $\frac{1}{3}$
14. $\frac{1}{6}$
15. $\frac{3}{14}$
16. $\frac{1}{10}$
17. $\frac{5}{8}$ inch

    $$\frac{\cancel{15}^5}{\cancel{16}_8} \times \frac{\cancel{2}^1}{\cancel{3}_1} = \frac{5}{8}$$

18. $\frac{1}{2}$ hour

    $$\frac{\cancel{3}^1}{\cancel{4}_2} \times \frac{\cancel{2}^1}{\cancel{3}_1} = \frac{1}{2}$$

Page 115

1. $\frac{8}{75}$
2. $\frac{1}{9}$

   Divide the numerator of the first fraction, 21, and the denominator of the third fraction, 9, by 3. Divide the new numerator of the first fraction, 7, and the denominator of the second fraction, 35, by 7. Cancel the numerator of the third fraction, 5, and the denominator of the second fraction, 5, by 5. Divide the numerator of the second fraction, 8, and the denominator of the first fraction, 24, by 8. Multiply the new numerators and denominators.

   $$\frac{21}{24} \times \frac{8}{35} \times \frac{5}{9} = \frac{\cancel{21}^1}{\cancel{24}_3} \times \frac{\cancel{8}^1}{\cancel{35}_1} \times \frac{\cancel{5}^1}{\cancel{9}_3} = \frac{1}{9}$$

3. $\frac{3}{5}$
4. $\frac{1}{15}$
5. $\frac{5}{84}$
6. $\frac{5}{18}$
7. $\frac{3}{5}$
8. $\frac{3}{112}$
9. $\frac{1}{180}$
10. $\frac{1}{2}$
11. $\frac{2}{27}$
12. $\frac{4}{7}$
13. $\frac{1}{4}$
14. $\frac{1}{12}$
15. $\frac{2}{45}$
16. $\frac{1}{3}$

17. $\frac{1}{40}$
18. $\frac{6}{35}$

Page 116

1. $\frac{3}{8}$
2. $\frac{1}{3}$
3. 1
4. $\frac{1}{48}$
5. $\frac{1}{2}$
6. $\frac{1}{4}$
7. $1\frac{1}{7}$
8. $\frac{3}{32}$
9. $\frac{1}{2}$
10. $\frac{1}{4}$
11. $\frac{8}{21}$
12. $\frac{1}{3}$
13. $\frac{10}{13}$
14. $\frac{2}{7}$
15. $\frac{1}{5}$
16. $\frac{4}{27}$
17. $\frac{1}{8}$
18. $\frac{1}{2}$
19. $\frac{3}{5}$
20. $\frac{4}{15}$
21. $\frac{1}{12}$
22. $\frac{1}{6}$
23. $\frac{1}{18}$
24. $\frac{1}{12}$
25. $\frac{4}{15}$
26. $\frac{3}{28}$
27. $\frac{1}{60}$
28. $\frac{1}{60}$
29. $\frac{1}{3}$
30. $\frac{1}{10}$
31. $\frac{1}{90}$
32. $\frac{1}{2}$
33. $\frac{4}{7}$
34. $\frac{3}{8}$ pound

$$\frac{13}{16} - \frac{7}{16} = \frac{6}{16} = \frac{3}{8}$$

35. $\frac{1}{3}$ teaspoon

$$\frac{1}{\cancel{2}} \times \frac{\cancel{2}^1}{3} = \frac{1}{3}$$

Page 117

1. 18 inches
2. 12 inches

   Set up the problem. Write 36 as an improper fraction with a denominator of 1. Cancel. Multiply the new fractions. The answer is an improper fraction. Change to a whole number by dividing the numerator by the denominator.

   $$\frac{1}{3} \times 36 = \frac{1}{\cancel{3}} \times \frac{\cancel{36}^{12}}{1} = \frac{12}{1} = 12$$

3. 27 inches
4. 24 inches
5. 12 hours
6. 8 hours
7. 18 hours
8. 6 days
9. 13 weeks
10. 39 weeks

Page 118

1. 10
2. 5

   Set up the problem. Write 10 as $\frac{10}{1}$. Cancel by dividing the numerator, 10, and the denominator, 2, by 2. Multiply the new numerators and denominators. The answer is an improper fraction. Change to a mixed number by dividing the numerator by the denominator.

   $$10 \times \frac{1}{2} = \frac{\cancel{10}^5}{1} \times \frac{1}{\cancel{2}_1} = \frac{5}{1} = 5$$

3. 6
4. 3
5. 4
6. 6
7. 3
8. 16
9. $5\frac{1}{2}$
10. $3\frac{7}{9}$
11. $9\frac{3}{8}$
12. $6\frac{2}{3}$
13. $1\frac{2}{7}$
14. $16\frac{1}{5}$
15. $16\frac{2}{3}$
16. $9\frac{1}{2}$
17. 6
18. 6
19. $\frac{1}{2}$
20. $\frac{5}{6}$

Page 119

1. $9\frac{1}{3}$
2. 22

   Write $4\frac{2}{5}$ as an improper fraction. Write 5 as an improper fraction. Cancel. Multiply. Write the answer as a whole number.

   $$4\frac{2}{5} \times 5 = \frac{22}{\cancel{5}} \times \frac{\cancel{5}^1}{1} = \frac{22}{1} = 22$$

3. $11\frac{1}{4}$
4. 23
5. 20
6. 55
7. 51
8. $19\frac{1}{2}$
9. $17\frac{1}{2}$
10. $31\frac{1}{3}$
11. $24\frac{4}{5}$
12. 40
13. $3\frac{2}{5}$
14. $14\frac{1}{3}$
15. $15\frac{3}{4}$
16. $19\frac{3}{5}$

17. 54

18. 21

19. $34\frac{4}{5}$

20. 50

21. 25 orders

$$10 \times 2\frac{1}{2} = \frac{\cancel{10}^{5}}{1} \times \frac{5}{\cancel{2}_{1}} = \frac{25}{1} = 25$$

22. $13\frac{1}{2}$ miles

$$2\frac{7}{10} \times 5 = \frac{27}{\cancel{10}_{2}} \times \frac{\cancel{5}^{1}}{1} = \frac{27}{2} = 13\frac{1}{2}$$

Page 120

1. $12 per hour

Marina's regular pay rate is $8 per hour. Multiply 8 by $1\frac{1}{2}$ to find her overtime rate.

$$8 \times 1\frac{1}{2} = \frac{\cancel{8}^{4}}{1} \times \frac{3}{\cancel{2}_{1}} = \frac{12}{1} = 12$$

2. $18 per hour

$$12 \times 1\frac{1}{2} = \frac{\cancel{12}^{6}}{1} \times \frac{3}{\cancel{2}_{1}} = \frac{18}{1} = 18$$

3. $15 per hour

$$10 \times 1\frac{1}{2} = \frac{\cancel{10}^{5}}{1} \times \frac{3}{\cancel{2}_{1}} = \frac{15}{1} = 15$$

4. $24 per hour

$$16 \times 1\frac{1}{2} = \frac{\cancel{16}^{8}}{1} \times \frac{3}{\cancel{2}_{1}} = \frac{24}{1} = 24$$

Marcy's Restaurant

Employee	Regular	Overtime
Kenichi	$ 6	$ 9
Marina	$ 8	$ 12
Nestor	$ 12	$ 18
Rita	$ 10	$ 15
Edward	$ 16	$ 24

Page 121

1. $21

Oscar's overtime pay rate is $6 per hour. Multiply 6 by $3\frac{1}{2}$ to find Oscar's overtime pay.

$$6 \times 3\frac{1}{2} = \frac{\cancel{6}^{3}}{1} \times \frac{7}{\cancel{2}_{1}} = \frac{21}{1} = 21$$

2. $51

$$9 \times 5\frac{2}{3} = \frac{\cancel{9}^{3}}{1} \times \frac{17}{\cancel{3}_{1}} = \frac{51}{1} = 51$$

3. $35

$$15 \times 2\frac{1}{3} = \frac{\cancel{15}^{5}}{1} \times \frac{7}{\cancel{3}_{1}} = \frac{35}{1} = 35$$

4. $153

$$18 \times 8\frac{1}{2} = \frac{\cancel{18}^{9}}{1} \times \frac{17}{\cancel{2}_{1}} = \frac{153}{1} = 153$$

EMPLOYEE	OVERTIME RATE	OVERTIME HOURS	OVERTIME PAY
Oscar	$ 6	$3\frac{1}{2}$	$ 21
Carl	$ 9	$5\frac{2}{3}$	$ 51
Ruthie	$ 15	$2\frac{1}{3}$	$ 35
James	$ 18	$8\frac{1}{2}$	$153
Mike	$ 12	$5\frac{3}{4}$	$ 69

Page 122

1. $2\frac{2}{5}$

2. $3\frac{1}{8}$

Write $4\frac{3}{8}$ as an improper fraction. Cancel. Multiply the numerators. Multiply the denominators. Change the answer to a mixed number.

$$4\frac{3}{8} \times \frac{5}{7} = \frac{\cancel{35}^{5}}{8} \times \frac{5}{\cancel{7}_{1}} = \frac{25}{8} = 3\frac{1}{8}$$

3. $1\frac{7}{20}$

4. $5\frac{1}{3}$

5. $\frac{3}{4}$

6. $\frac{1}{4}$

7. $3\frac{4}{5}$

8. $\frac{23}{49}$

9. $1\frac{9}{16}$

10. $2\frac{5}{6}$

11. $5\frac{1}{10}$

12. $3\frac{12}{13}$

13. $4\frac{5}{6}$

14. $2\frac{11}{16}$

15. $4\frac{5}{24}$

16. $1\frac{14}{15}$

17. $1\frac{11}{27}$

18. $1\frac{9}{11}$

Page 123

1. $16\frac{1}{3}$

2. 35

Write the mixed numbers as improper fractions. Cancel. Multiply. Change the answer to a whole number.

$$9\frac{4}{5} \times 3\frac{4}{7} = \frac{\cancel{49}^7}{\cancel{5}_1} \times \frac{\cancel{25}^5}{\cancel{7}_1} = \frac{35}{1} = 35$$

3. $45\frac{1}{2}$ **4.** $6\frac{3}{4}$

5. 14   **6.** $3\frac{2}{5}$

7. $32\frac{2}{3}$ **8.** 11

9. $13\frac{2}{7}$ **10.** $4\frac{1}{9}$

11. $10\frac{4}{5}$ **12.** $3\frac{9}{10}$

13. $47\frac{1}{2}$ **14.** $22\frac{10}{11}$

15. 78

Page 124

1. $5\frac{1}{4}$ ounces

$$1\frac{1}{2} \times 3\frac{1}{2} = \frac{3}{2} \times \frac{7}{2} = \frac{21}{4} = 5\frac{1}{4}$$

2. $4\frac{3}{8}$ cups

$$1\frac{1}{4} \times 3\frac{1}{2} = \frac{5}{4} \times \frac{7}{2} = \frac{35}{8} = 4\frac{3}{8}$$

3. $2\frac{5}{8}$ cups

$$\frac{3}{4} \times 3\frac{1}{2} = \frac{3}{4} \times \frac{7}{2} = \frac{21}{8} = 2\frac{5}{8}$$

4. $\frac{7}{16}$ teaspoon

$$\frac{1}{8} \times 3\frac{1}{2} = \frac{1}{8} \times \frac{7}{2} = \frac{7}{16}$$

5. $2\frac{1}{2}$ cups

$$\frac{1}{2} \times 5 = \frac{1}{2} \times \frac{5}{1} = \frac{5}{2} = 2\frac{1}{2}$$

6. $3\frac{3}{4}$ cups

$$\frac{3}{4} \times 5 = \frac{3}{4} \times \frac{5}{1} = \frac{15}{4} = 3\frac{3}{4}$$

Unit 4 Review, page 125

1. $\frac{2}{9}$ **2.** $\frac{1}{6}$

3. $\frac{1}{12}$ **4.** $\frac{3}{10}$

5. $\frac{5}{36}$ **6.** $\frac{16}{25}$

7. $\frac{35}{48}$ **8.** $\frac{7}{20}$

9. $\frac{1}{24}$ **10.** $\frac{9}{140}$

11. $\frac{3}{80}$ **12.** $\frac{1}{4}$

13. $\frac{1}{42}$ **14.** $\frac{4}{45}$

15. $\frac{1}{7}$ **16.** $\frac{2}{3}$

17. $\frac{1}{6}$ **18.** $\frac{1}{9}$

19. $\frac{3}{20}$ **20.** $\frac{3}{35}$

21. $\frac{1}{4}$ **22.** $\frac{1}{7}$

23. $\frac{1}{10}$ **24.** $\frac{2}{9}$

25. $\frac{1}{6}$ **26.** $\frac{1}{9}$

27. $\frac{1}{5}$ **28.** $\frac{3}{10}$

29. $\frac{2}{45}$ **30.** $\frac{1}{15}$

31. $\frac{5}{84}$ **32.** $\frac{5}{18}$

33. $\frac{3}{32}$ **34.** $\frac{1}{9}$

35. $\frac{7}{60}$

Page 126

36. 5   **37.** 8

38. 15   **39.** 24

40. $3\frac{1}{3}$ **41.** $1\frac{3}{4}$

42. $3\frac{3}{5}$ **43.** $5\frac{1}{2}$

44. 8   **45.** $2\frac{2}{3}$

46. 7   **47.** 10

48. $4\frac{1}{2}$ **49.** 34

50. $9\frac{1}{2}$ **51.** $34\frac{5}{7}$

52. $1\frac{2}{5}$ **53.** $1\frac{8}{27}$

54. $2\frac{11}{30}$ **55.** $1\frac{3}{11}$

56. $3\frac{1}{6}$ **57.** 14

58. $11\frac{1}{7}$ **59.** 8

Unit 5

Page 127

1. $\frac{2}{9}$ **2.** $\frac{4}{7}$

$$\frac{2}{\cancel{3}_1} \times \frac{\cancel{6}^3}{7} = \frac{4}{7}$$

3. $\frac{1}{2}$ **4.** $\frac{2}{3}$

198

5. $\frac{1}{12}$
6. $\frac{1}{14}$
7. $\frac{11}{36}$
8. $\frac{3}{7}$
9. $\frac{4}{35}$
10. $\frac{4}{7}$

Page 128

11. 6
12. $4\frac{1}{4}$

$$17 \times \frac{1}{4} = \frac{17}{1} \times \frac{1}{4} = \frac{17}{4} = 4\frac{1}{4}$$

13. 30
14. 4
15. 9
16. 2
17. $\frac{4}{5}$
18. $3\frac{1}{9}$
19. $5\frac{1}{6}$
20. $2\frac{13}{14}$
21. 5
22. $2\frac{3}{5}$

$$\begin{array}{r} 2\frac{3}{5} \\ 5\overline{)13} \\ -10 \\ \hline 3 \end{array}$$

23. $5\frac{2}{3}$
24. $3\frac{3}{4}$
25. 4
26. $5\frac{1}{6}$
27. $2\frac{1}{4}$
28. $1\frac{1}{2}$
29. 2
30. 4
31. $\frac{1}{2}$
32. LT
3 and 8 cannot be divided evenly by any other number but 1.
33. $\frac{1}{3}$
34. $\frac{1}{2}$
35. LT
36. $\frac{2}{5}$
37. $\frac{2}{3}$
38. $\frac{3}{4}$
39. $\frac{3}{7}$
40. $\frac{2}{3}$
41. $\frac{3}{2}$
42. $\frac{8}{3}$

Multiply the denominator, 3, by the whole number, 2. $3 \times 2 = 6$. Add 6 to the numerator. $6 + 2 = 8$. Write 8 over the denominator, 3.

43. $\frac{17}{4}$
44. $\frac{17}{5}$
45. $\frac{43}{8}$

Page 129

1. 5
2. 6

Invert the fraction to the right of the division sign. Change the division sign to a multiplication sign. Multiply by the new fraction. Cancel. Change the answer to a whole number.

$$\frac{3}{4} \div \frac{1}{8} = \frac{3}{\cancel{4}} \times \frac{\cancel{8}^2}{1} = \frac{6}{1} = 6$$

3. 4
4. 2
5. 6
6. 14
7. 6
8. 9
9. 10
10. 4

Page 130

1. $1\frac{1}{4}$
2. $2\frac{2}{9}$

Invert the fraction to the right of the division sign. Multiply. Change the answer to a mixed number.

$$\frac{5}{9} \div \frac{1}{4} = \frac{5}{9} \times \frac{4}{1} = \frac{20}{9} = 2\frac{2}{9}$$

3. $1\frac{11}{21}$
4. $1\frac{1}{5}$
5. $\frac{1}{2}$
6. $\frac{18}{25}$

Invert the fraction to the right of the division sign. Multiply. The answer is a fraction in lowest terms.

$$\frac{2}{5} \div \frac{5}{9} = \frac{2}{5} \times \frac{9}{5} = \frac{18}{25}$$

7. $\frac{3}{4}$
8. $\frac{5}{7}$
9. $\frac{8}{9}$
10. $2\frac{1}{4}$
11. $\frac{5}{6}$
12. $4\frac{2}{3}$
13. $1\frac{9}{16}$
14. $\frac{22}{25}$
15. $2\frac{1}{2}$
16. $\frac{1}{2}$
17. 2 pieces

Divide $\frac{1}{2}$ by $\frac{1}{4}$. Invert the fraction to the right of the division sign. Cancel and multiply. Change the answer to a whole number.

$$\frac{1}{2} \div \frac{1}{4} = \frac{1}{\cancel{2}} \times \frac{\cancel{4}^2}{1} = \frac{2}{1} = 2$$

18. 3 pieces
Divide $\frac{1}{2}$ by $\frac{1}{6}$.

$$\frac{1}{2} \div \frac{1}{6} = \frac{1}{\cancel{2}} \times \frac{\cancel{6}^3}{1} = \frac{3}{1} = 3$$

Page 131

1. 4
2. 3

   Change 1 to $\frac{1}{1}$. Invert $\frac{1}{3}$. Multiply the new fractions. Change the answer to a whole number.

   $$1 \div \frac{1}{3} = \frac{1}{1} \times \frac{3}{1} = \frac{3}{1} = 3$$

3. 8
4. 18
5. 16
6. 25
7. 20
8. 24
9. 54
10. 14
11. 45
12. 48
13. 100
14. 40

Page 132

1. $7\frac{1}{2}$
2. $6\frac{2}{3}$

   Change 5 to $\frac{5}{1}$. Invert $\frac{3}{4}$. Multiply. Change to a mixed number.

   $$5 \div \frac{3}{4} = \frac{5}{1} \times \frac{4}{3} = \frac{20}{3} = 6\frac{2}{3}$$

3. 15
4. $10\frac{4}{5}$
5. 28
6. $15\frac{3}{4}$
7. 25
8. $14\frac{6}{7}$
9. 20
10. 60
11. 81
12. 33
13. 21
14. $9\frac{3}{5}$
15. 50
16. $17\frac{1}{3}$
17. 24 pieces
    Divide 2 by $\frac{1}{12}$.

    $$2 \div \frac{1}{12} = \frac{2}{1} \times \frac{12}{1} = \frac{24}{1} = 24$$

18. 18 pieces
    Divide 2 by $\frac{1}{9}$.

    $$2 \div \frac{1}{9} = \frac{2}{1} \times \frac{9}{1} = \frac{18}{1} = 18$$

Page 133

1. 4
2. 10

   Change the mixed number to an improper fraction. Invert and multiply. Cancel. Change the answer to a whole number.

   $$2\frac{1}{2} \div \frac{1}{4} = \frac{5}{\cancel{2}} \times \frac{\cancel{4}^2}{1} = \frac{10}{1} = 10$$

3. 14
4. 15
5. 21
6. 10
7. 23
8. 68
9. 49
10. 126
11. 36

Page 134

1. $6\frac{3}{4}$
2. $43\frac{1}{5}$

   Change $5\frac{2}{5}$ to an improper fraction. Invert $\frac{1}{8}$ and multiply. Change the answer to a mixed number.

   $$5\frac{2}{5} \div \frac{1}{8} = \frac{27}{5} \times \frac{8}{1} = \frac{216}{5} = 43\frac{1}{5}$$

3. $3\frac{5}{6}$
4. $18\frac{1}{3}$
5. $46\frac{2}{7}$
6. $48\frac{3}{5}$
7. $17\frac{1}{2}$
8. $2\frac{1}{32}$
9. $5\frac{7}{22}$
10. $31\frac{3}{4}$
11. $46\frac{1}{5}$
12. $153\frac{1}{2}$
13. $4\frac{2}{3}$
14. $9\frac{1}{6}$
15. $7\frac{5}{6}$
16. $3\frac{1}{45}$
17. $4\frac{2}{3}$ loaves
    Divide $3\frac{1}{2}$ by $\frac{3}{4}$.

    $$3\frac{1}{2} \div \frac{3}{4} = \frac{7}{2} \div \frac{3}{4} = \frac{7}{\cancel{2}} \times \frac{\cancel{4}^2}{3} = \frac{14}{3} = 2\frac{2}{3}$$

18. $10\frac{2}{3}$ dozen
    Divide $5\frac{1}{3}$ by $\frac{1}{2}$.

    $$5\frac{1}{3} \div \frac{1}{2} = \frac{16}{3} \div \frac{1}{2} = \frac{16}{3} \times \frac{2}{1} = \frac{32}{3} = 10\frac{2}{3}$$

Page 135

1. 20 pieces

   $$10 \div \frac{1}{2} = \frac{10}{1} \times \frac{2}{1} = \frac{20}{1} = 20$$

2. 72 packages

   $$18 \div \frac{1}{4} = \frac{18}{1} \times \frac{4}{1} = \frac{72}{1} = 72$$

3. 8 packages

$$\frac{1}{2} \div \frac{1}{16} = \frac{1}{\cancel{2}} \times \frac{\cancel{16}^{8}}{1} = \frac{8}{1} = 8$$

4. 8 packages

$$1 \div \frac{1}{8} = \frac{1}{1} \div \frac{1}{8} = \frac{1}{1} \times \frac{8}{1} = \frac{8}{1} = 8$$

5. 7 packages; $\frac{7}{20}$ pound is left.

$$4\frac{9}{10} \div \frac{2}{3} = \frac{49}{10} \div \frac{2}{3} = \frac{49}{10} \times \frac{3}{2} = \frac{147}{20} = 7\frac{7}{20}$$

6. 34 packages; $\frac{1}{3}$ pound is left.

$$25\frac{3}{4} \div \frac{3}{4} = \frac{103}{4} \div \frac{3}{4} = \frac{103}{\cancel{4}_1} \times \frac{\cancel{4}^1}{3} = \frac{103}{3} = 34\frac{1}{3}$$

Page 136

1. 4
2. $\frac{5}{6}$
3. $\frac{3}{20}$
4. 4
5. $1\frac{1}{2}$
6. $\frac{19}{56}$
7. $\frac{1}{12}$
8. $\frac{1}{3}$
9. $\frac{1}{9}$
10. 1
11. $\frac{53}{60}$
12. $\frac{3}{8}$
13. $2\frac{2}{3}$
14. 6
15. $2\frac{8}{9}$
16. 27
17. $6\frac{1}{5}$
18. $6\frac{3}{7}$
19. $6\frac{2}{5}$
20. $8\frac{1}{3}$
21. 32
22. $17\frac{1}{7}$
23. $5\frac{1}{2}$
24. 48
25. 21
26. 2
27. $1\frac{5}{9}$
28. 38
29. $2\frac{19}{32}$
30. $1\frac{5}{8}$
31. $12\frac{1}{3}$
32. $32\frac{8}{15}$
33. 12
34. $2\frac{2}{7}$
35. $27\frac{13}{15}$
36. $6\frac{7}{18}$

Page 137

1. $\frac{1}{8}$

2. $\frac{3}{10}$

Change 2 to an improper fraction. Invert. Multiply.

$$\frac{3}{5} \div 2 = \frac{3}{5} \div \frac{2}{1} = \frac{3}{5} \times \frac{1}{2} = \frac{3}{10}$$

3. $\frac{1}{6}$
4. $\frac{1}{10}$
5. $\frac{1}{14}$
6. $\frac{1}{48}$
7. $\frac{1}{10}$
8. $\frac{2}{63}$
9. $\frac{1}{8}$
10. $\frac{7}{30}$
11. $\frac{1}{18}$
12. $\frac{5}{18}$
13. $\frac{4}{25}$
14. $\frac{1}{36}$

Page 138

1. $\frac{1}{3}$

2. $\frac{1}{8}$

Change 6 to an improper fraction. Invert and multiply. Cancel.

$$\frac{3}{4} \div 6 = \frac{3}{4} \div \frac{6}{1} = \frac{\cancel{3}^1}{4} \times \frac{1}{\cancel{6}_2} = \frac{1}{8}$$

3. $\frac{7}{40}$
4. $\frac{1}{18}$
5. $\frac{1}{7}$
6. $\frac{9}{40}$
7. $\frac{1}{24}$
8. $\frac{3}{16}$
9. $\frac{1}{11}$
10. $\frac{1}{20}$
11. $\frac{1}{42}$
12. $\frac{1}{24}$

13. $\frac{3}{8}$ pound

$$\frac{3}{4} \div 2 = \frac{3}{4} \div \frac{2}{1} = \frac{3}{4} \times \frac{1}{2} = \frac{3}{8}$$

14. $\frac{7}{24}$ yard

$$\frac{7}{8} \div 3 = \frac{7}{8} \div \frac{3}{1} = \frac{7}{8} \times \frac{1}{3} = \frac{7}{24}$$

Page 139

1. $2\frac{5}{24}$

2. $2\frac{5}{6}$

Change $5\frac{2}{3}$ to an improper fraction.

Change 2 to an improper fraction. Invert and multiply. Change to a mixed number.

$$5\frac{2}{3} \div 2 = \frac{17}{3} \div \frac{2}{1} = \frac{17}{3} \times \frac{1}{2} = \frac{17}{6} = 2\frac{5}{6}$$

3. $1\frac{16}{25}$
4. $1\frac{5}{21}$

5. $\frac{7}{8}$
6. $\frac{17}{42}$
7. $\frac{11}{18}$
8. $\frac{23}{24}$
9. $2\frac{9}{55}$
10. $1\frac{29}{96}$
11. $5\frac{1}{4}$ weeks

$10\frac{1}{2} \div 2 = \frac{21}{2} \div \frac{2}{1} = \frac{21}{2} \times \frac{1}{2} = \frac{21}{4} = 5\frac{1}{4}$

12. $4\frac{1}{4}$ weeks

$12\frac{3}{4} \div 3 = \frac{51}{4} \div \frac{3}{1} = \frac{\overset{17}{\cancel{51}}}{4} \times \frac{1}{\underset{1}{\cancel{3}}} = \frac{17}{4} = 4\frac{1}{4}$

Page 140

1. $3\frac{5}{7}$
2. $1\frac{15}{31}$

   Change both mixed numbers to improper fractions. Invert and multiply. Cancel. Change to a mixed number.

   $7\frac{2}{3} \div 5\frac{1}{6} =$

   $\frac{23}{3} \div \frac{31}{6} = \frac{23}{\underset{1}{\cancel{3}}} \times \frac{\overset{2}{\cancel{6}}}{31} = \frac{46}{31} = 1\frac{5}{31}$

3. $2\frac{20}{33}$
4. $4\frac{2}{9}$
5. $\frac{85}{93}$
6. $\frac{60}{71}$
7. $\frac{48}{55}$
8. $5\frac{19}{21}$
9. 1
10. $1\frac{34}{35}$
11. 5 windows

    $18\frac{3}{4} \div 3\frac{3}{4} =$

    $\frac{75}{4} \div \frac{15}{4} = \frac{\overset{5}{\cancel{75}}}{\underset{1}{\cancel{4}}} \times \frac{\overset{1}{\cancel{4}}}{\underset{1}{\cancel{15}}} = \frac{5}{1} = 5$

12. 3 laps

    $7\frac{1}{2} \div 2\frac{1}{2} = \frac{15}{2} \div \frac{5}{2} = \frac{\overset{3}{\cancel{15}}}{\underset{1}{\cancel{2}}} \times \frac{\overset{1}{\cancel{2}}}{\underset{1}{\cancel{5}}} = \frac{3}{1} = 3$

Page 141

1. 4 shelves

   48 inches $\div$ $11\frac{1}{2}$ inches $=$

   $\frac{48}{1} \div \frac{23}{2} = \frac{48}{1} \times \frac{2}{23} = \frac{96}{23} = 4\frac{2}{43}$

   Hector will be able to put 4 whole shelves in the bookcase.

2. $\frac{5}{6}$ foot apart

   $6\frac{2}{3}$ feet $\div 8 = \frac{20}{3} \div \frac{8}{1} = \frac{\overset{5}{\cancel{20}}}{3} \times \frac{1}{\underset{2}{\cancel{8}}} = \frac{5}{6}$

Page 142

3. 3 tops

   12 feet $\div$ $3\frac{1}{2}$ feet $=$

   $\frac{12}{1} \div \frac{7}{2} = \frac{12}{1} \times \frac{2}{7} = \frac{24}{7} = 3\frac{3}{7}$

   He can cut 3 whole bookcase tops from one 12-foot board.

4. 2 sides

   14 feet $\div$ $6\frac{2}{3}$ feet $=$

   $\frac{14}{1} \div \frac{20}{3} = \frac{\overset{7}{\cancel{14}}}{1} \times \frac{3}{\underset{10}{\cancel{20}}} = \frac{21}{10} = 2\frac{1}{10}$

   He can cut 2 complete sides from one 14-foot board.

5. 7 shelves

   24 feet $\div$ $3\frac{5}{12}$ feet $=$

   $\frac{24}{1} \div \frac{41}{12} = \frac{24}{1} \times \frac{12}{41} = \frac{288}{41} = 7\frac{1}{41}$

   He can cut 7 whole shelves from one 24-foot board.

6. $1\frac{3}{4}$ feet

   $3\frac{1}{2}$ feet $\div 2 =$

   $\frac{7}{2} \div \frac{2}{1} = \frac{7}{2} \times \frac{1}{2} = \frac{7}{4} = 1\frac{3}{4}$

Unit 5 Review, page 143

1. $1\frac{1}{3}$
2. $\frac{1}{2}$
3. $\frac{9}{14}$
4. $1\frac{1}{9}$
5. $1\frac{5}{22}$
6. $3\frac{3}{4}$
7. 4
8. $\frac{1}{2}$
9. 8
10. 27
11. $9\frac{1}{3}$
12. $12\frac{1}{2}$
13. $21\frac{1}{3}$
14. 21
15. $14\frac{2}{5}$
16. 12
17. 11
18. $17\frac{1}{2}$
19. $9\frac{5}{6}$
20. $5\frac{3}{5}$
21. $29\frac{1}{2}$
22. $17\frac{1}{2}$
23. 107
24. $9\frac{1}{5}$

25. $\frac{1}{27}$
26. $\frac{3}{8}$
27. $\frac{1}{8}$
28. $\frac{1}{9}$
29. $\frac{9}{100}$
30. $\frac{5}{24}$
31. $\frac{1}{24}$
32. $\frac{1}{14}$
33. $\frac{2}{25}$
34. $\frac{2}{27}$
35. $\frac{4}{19}$
36. $\frac{3}{32}$

Page 144

37. $1\frac{3}{8}$
38. $\frac{1}{5}$
39. $2\frac{23}{27}$
40. $\frac{11}{15}$
41. $3\frac{11}{16}$
42. $1\frac{1}{20}$
43. $\frac{39}{40}$
44. $\frac{25}{48}$
45. $\frac{62}{123}$
46. $2\frac{19}{39}$
47. $1\frac{88}{185}$
48. $2\frac{1}{3}$
49. $4\frac{1}{33}$
50. $6\frac{1}{15}$
51. $8\frac{13}{16}$
52. $4\frac{17}{30}$
53. $1\frac{9}{13}$

Unit 6

Page 145

1. $1\frac{5}{12}$
2. $2\frac{1}{8}$

$$1\frac{5}{8}$$
$$+\ \frac{4}{8}$$
$$1\frac{9}{8} = 2\frac{1}{8}$$

3. $14\frac{1}{6}$
4. $1\frac{31}{40}$
5. 5
6. 1
7. $3\frac{7}{20}$
8. $3\frac{1}{3}$
9. $11\frac{1}{8}$
10. $3\frac{2}{9}$

Page 146

11. $\frac{1}{2}$
12. $1\frac{1}{2}$

$$1\frac{2}{3} = 1\frac{4}{6}$$
$$-\ \frac{1}{6} = \frac{1}{6}$$
$$1\frac{3}{6} = 1\frac{1}{2}$$

13. $2\frac{3}{10}$
14. $\frac{5}{8}$
15. $\frac{3}{14}$
16. $\frac{3}{4}$
17. $6\frac{7}{12}$
18. $5\frac{3}{10}$
19. $1\frac{4}{5}$
20. $\frac{7}{20}$
21. $\frac{3}{10}$
22. $1\frac{2}{9}$

$$1\frac{3}{8} \times \frac{8}{9} = \frac{11}{\cancel{8}} \times \frac{\cancel{8}^1}{9} = \frac{11}{9} = 1\frac{2}{9}$$

23. $9\frac{2}{3}$
24. 3
25. $\frac{1}{6}$
26. $3\frac{5}{12}$
27. $20\frac{1}{2}$
28. $\frac{5}{24}$
29. 5
30. 40

$$8 \div \frac{1}{5} = \frac{8}{1} \times \frac{5}{1} = \frac{40}{1} = 40$$

31. $2\frac{5}{11}$
32. $\frac{51}{104}$
33. $3\frac{3}{5}$
34. $\frac{8}{15}$
35. $4\frac{2}{3}$
36. $1\frac{49}{51}$

Page 147

1. 18 inches

$$1\frac{1}{2} \times 12 = \frac{3}{\cancel{2}_1} \times \frac{\cancel{12}^6}{1} = \frac{18}{1} = 18$$

2. 440 yards

$$\frac{1}{4} \times 1{,}760 = \frac{1}{\cancel{4}_1} \times \frac{\cancel{1760}^{440}}{1} = \frac{440}{1} = 440$$

3. c, 11,800 feet

$$2\frac{1}{4} \times 5{,}280 = \frac{9}{\cancel{4}_1} \times \frac{\cancel{5280}^{1320}}{1} = \frac{11800}{1} = 11{,}800$$

4. b, 3,960 yards

$$2\frac{1}{4} \times 1{,}760 = \frac{9}{\cancel{4}_1} \times \frac{\cancel{1760}^{440}}{1} = \frac{3960}{1} = 3{,}960$$

Page 148

1. Crispy Cherry Bites cost more.
   $\frac{79}{10} = 7\frac{9}{10}$ cents
   $\frac{89}{11} = 8\frac{1}{11}$ cents
   $8\frac{1}{11} > 7\frac{9}{10}$

2. Red apples cost less.
   $\frac{99}{4} = 24\frac{3}{4}$ cents
   $\frac{69}{3} = 23$ cents
   $23 < 24\frac{3}{4}$

3. Meaty hot dogs cost less.
   $\frac{6}{7} = \frac{36}{42}$
   $\frac{5}{6} = \frac{35}{42}$
   $\frac{35}{42} < \frac{36}{42}$

4. Squeezy costs more.
   $\frac{2}{3} = \frac{16}{24}$
   $\frac{6}{8} = \frac{18}{24}$
   $\frac{18}{24} > \frac{16}{24}$

Page 149

1. $1\frac{3}{4}$ points

   $13\frac{1}{4} = 13\frac{1}{4} = 12\frac{5}{4}$
   $- 11\frac{1}{2} = 11\frac{2}{4} = 11\frac{2}{4}$
   $\phantom{-11\frac{1}{2} = 11\frac{2}{4} = 11} 1\frac{3}{4}$

2. $1\frac{1}{2}$ points

   $12\frac{3}{8} = 11\frac{11}{8}$
   $- 10\frac{7}{8} = 10\frac{7}{8}$
   $\phantom{- 10\frac{7}{8} = 10} 1\frac{4}{8} = 1\frac{1}{2}$

3. c, $265

   $20 \times 13\frac{1}{4} = \frac{\overset{5}{\cancel{20}}}{1} \times \frac{53}{\underset{1}{\cancel{4}}} = \$265$

4. a, $15\frac{7}{8}$

   $14\frac{3}{4} = 14\frac{6}{8}$
   $+ 1\frac{1}{8} = 1\frac{1}{8}$
   $\phantom{+ 1\frac{1}{8} = } 15\frac{7}{8}$

Page 150

1. He needs 1 more quart.
   $\frac{1}{2} \times 4 = \frac{1}{\underset{1}{\cancel{2}}} \times \frac{\overset{2}{\cancel{4}}}{1} = \frac{2}{1} = 2$ quarts
   2 quarts $- 1$ quart $= 1$ quart

2. He needs 2 more quarts.
   $3\frac{1}{2} \times 4 = \frac{7}{\underset{1}{\cancel{2}}} \times \frac{\overset{2}{\cancel{4}}}{1} = \frac{14}{1} = 14$ quarts
   $14 - 12 = 2$ quarts

3. Dawn needs 2 more cups.
   $\frac{1}{2} \times 2 = \frac{1}{\underset{1}{\cancel{2}}} \times \frac{\overset{1}{\cancel{2}}}{1} = \frac{1}{1} = 1$ cup
   $3 - 1 = 2$ cups

4. 3 quarts
   $\frac{1}{2} \times 4 = \frac{1}{\underset{1}{\cancel{2}}} \times \frac{\overset{2}{\cancel{4}}}{1} = \frac{2}{1} = 2$ quarts
   $2 + 1 = 3$ quarts

Page 151

5. 22 pints
   $20 \times \frac{1}{2} = \frac{\overset{10}{\cancel{20}}}{1} \times \frac{1}{\underset{1}{\cancel{2}}} = \frac{10}{1} = 10$ pints
   $12 + 10 = 22$ pints

6. $1\frac{1}{2}$ cups
   $24 \div 8 = 3$ cups
   $ 3 = 2\frac{2}{2}$
   $- 1\frac{1}{2} = 1\frac{1}{2}$
   $\phantom{- 1\frac{1}{2} = } 1\frac{1}{2}$ cups

7. 36 ounces
   $3 \times 1\frac{1}{2} = \frac{3}{1} \times \frac{3}{2} = \frac{9}{2} = 4\frac{1}{2}$ cups
   $4\frac{1}{2} \times 8 = \frac{9}{\underset{1}{\cancel{2}}} \times \frac{\overset{4}{\cancel{8}}}{1} = \frac{36}{1} = 36$ ounces

8. $\frac{1}{4}$ cup or 2 ounces

   $2 \times 8 = 16$ ounces

   $16 \div 8 = 2$ cups

   $2 = 1\frac{4}{4}$
   $\underline{-1\frac{3}{4} = 1\frac{3}{4}}$
   $\phantom{-1\frac{3}{4} = 1}\frac{1}{4}$ cup

   or

   $1\frac{3}{4} \times 8 = \frac{7}{\cancel{4}} \times \frac{\cancel{8}^2}{1} = \frac{14}{1} = 14$ ounces

   $16 - 14 = 2$ ounces

9. 8 servings

   $32 \div 8 = 4$ cups

   $4 \div \frac{1}{2} = \frac{4}{1} \div \frac{1}{2} = \frac{4}{1} \times \frac{2}{1} = \frac{8}{1} = 8$

10. 6 rooms

    $2\frac{1}{2} \times 4 = \frac{5}{\cancel{2}} \times \frac{\cancel{4}^2}{1} = \frac{10}{1} = 10$ quarts

    $10 \div 1\frac{2}{3} = \frac{10}{1} \div \frac{5}{3} = \frac{\cancel{10}^2}{1} \times \frac{3}{\cancel{5}} = \frac{6}{1} = 6$

11. 11 pots of coffee

    $22 \div 8 = \frac{22}{1} \div \frac{8}{1} = \frac{\cancel{22}^{11}}{1} \times \frac{1}{\cancel{8}_4} = \frac{11}{4} =$

    $2\frac{3}{4}$ cups

    $2\frac{3}{4} \div \frac{1}{4} = \frac{11}{\cancel{4}} \times \frac{\cancel{4}^1}{1} = \frac{11}{1} = 11$

12. 12 servings

    $1\frac{1}{2}$ quarts $\times 2 = \frac{3}{\cancel{2}} \times \frac{\cancel{2}^1}{1} = 3$ pints

    3 pints $\times$ 2 cups = 6 cups

    $6 \div \frac{1}{2} = \frac{6}{1} \times \frac{2}{1} = \frac{12}{1} = 12$

Page 153

1. 10 pounds

   $\frac{20}{4} = \frac{?}{2}$

   $\frac{20}{4} = \frac{20 \div 2}{4 \div 2} = \frac{10}{2}$

2. 2 pounds

   $\frac{4}{2} = \frac{?}{1}$

   $\frac{4}{2} = \frac{4 \div 2}{2 \div 2} = \frac{2}{1}$

3. $6

   $\frac{2}{3} = \frac{4}{?}$

   $\frac{2}{3} = \frac{2 \times 2}{3 \times 2} = \frac{4}{6}$

4. $15

   $\frac{2}{5} = \frac{6}{?}$

   $\frac{2}{5} = \frac{2 \times 3}{5 \times 3} = \frac{6}{15}$

5. $12

   $\frac{5}{4} = \frac{15}{?}$

   $\frac{5}{4} = \frac{5 \times 3}{4 \times 3} = \frac{15}{12}$

6. 6 pies

   $\frac{3}{10} = \frac{?}{20}$

   $\frac{3}{10} = \frac{3 \times 2}{10 \times 2} = \frac{6}{20}$

7. 4 pounds

   $\frac{1}{2} = \frac{8}{?}$

   $\frac{1}{2} = \frac{1 \times 4}{2 \times 4} = \frac{4}{8}$

8. $10

   $\frac{3}{5} = \frac{6}{?}$

   $\frac{3}{5} = \frac{3 \times 2}{5 \times 2} = \frac{6}{10}$

9. $2

   $\frac{6}{4} = \frac{3}{?}$

   $\frac{6}{4} = \frac{6 \times 2}{4 \times 2} = \frac{2}{1}$

10. 9 pints

    $\frac{3}{4} = \frac{?}{13}$

    $\frac{3}{4} = \frac{3 \div 3}{4 \div 3} = \frac{9}{12}$

11. $2

    $\frac{3}{1} = \frac{6}{?}$

    $\frac{3}{1} = \frac{3 \times 2}{1 \times 2} = \frac{6}{2}$

12. $6

    $3 \times 12 = 36$

    $\frac{6}{1} = \frac{36}{?}$

    $\frac{6}{1} = \frac{6 \times 6}{1 \times 6} = \frac{36}{6}$

Page 155

1. $3\frac{1}{2}$ hours

    11 hours 30 minutes
    − 8 hours 0 minutes
    3 hours 30 minutes = $3\frac{30}{60} = 3\frac{1}{2}$

2. $4\frac{1}{2}$ hours

    12 hours 30 minutes
    − 8 hours 0 minutes
    4 hours 30 minutes = $4\frac{30}{60} = 4\frac{1}{2}$

3. $2\frac{1}{4}$ hours

    3 hours 45 minutes
    − 1 hour 30 minutes
    2 hours 15 minutes = $2\frac{15}{60} = 2\frac{1}{4}$

4. $5\frac{1}{2}$ hours

    11 hours 30 minutes
    − 6 hours 0 minutes
    5 hours 30 minutes = $5\frac{30}{60} = 5\frac{1}{2}$

5. $3\frac{1}{2}$ hours

    6 hours 45 minutes
    − 3 hours 15 minutes
    3 hours 30 minutes = $3\frac{30}{60} = 3\frac{1}{2}$

6. 3 hours

    6 hours 30 minutes
    − 3 hours 30 minutes
    3 hours 0 minutes

7. $17

    $4 \times 4\frac{1}{4} = \frac{\cancel{4}^1}{1} \times \frac{17}{\cancel{4}_1} = \frac{17}{1} = 17

8. $2\frac{3}{4}$ hours, $11

    3 hours 45 minutes
    − 1 hour 0 minutes
    2 hours 45 minutes = $2\frac{45}{60} = 2\frac{3}{4}$

    $4 \times 2\frac{3}{4} = \frac{\cancel{4}^1}{1} \times \frac{11}{\cancel{4}_1} = \frac{11}{1} = 11

Child	Mon.	Tues.	Wed.	Thurs.	Fri.
Teddy	7:30-11:45	8-11:30	Sick	Sick	8-12:30
Peter H.	1-3:45	12:30-6:30	1:15-5:30	2-6:30	—
Susie	5:30-10:45	—	6-11:30	—	5:45-10:45
Peter W.	—	—	—	1:30-3:45	—
Hannah	3:15-6:45	—	—	—	3:30-6:30

Skills Inventory

Page 156

1. $\frac{1}{2}$
2. $\frac{1}{4}$
3. LT
4. $\frac{3}{4}$
5. $\frac{2}{3}$
6. $\frac{4}{12}$
7. $\frac{6}{21}$
8. $\frac{5}{10}$
9. $\frac{16}{20}$
10. $\frac{20}{32}$
11. $1\frac{1}{4}$
12. 2
13. $2\frac{1}{7}$
14. 4
15. $2\frac{1}{10}$
16. $\frac{12}{5}$
17. $\frac{10}{3}$
18. $\frac{23}{4}$
19. $\frac{11}{9}$
20. $\frac{97}{12}$
21. $\frac{1}{2}$
22. 1
23. $1\frac{1}{2}$
24. $1\frac{1}{8}$
25. $4\frac{6}{7}$
26. $10\frac{2}{5}$
27. $13\frac{1}{3}$
28. 6
29. $7\frac{3}{4}$
30. $\frac{1}{2}$
31. $\frac{17}{20}$
32. $\frac{5}{7}$
33. $\frac{37}{40}$
34. $1\frac{11}{36}$

Page 157

35. $8\frac{1}{21}$
36. $10\frac{3}{8}$
37. $14\frac{1}{7}$
38. $15\frac{19}{20}$
39. $5\frac{23}{24}$
40. $\frac{1}{2}$
41. $\frac{4}{9}$
42. 0
43. $\frac{3}{5}$
44. $6\frac{2}{3}$
45. $6\frac{2}{5}$
46. $4\frac{1}{2}$
47. $2\frac{2}{5}$
48. $6\frac{10}{21}$
49. $\frac{5}{7}$
50. $4\frac{7}{9}$
51. $6\frac{2}{5}$
52. $1\frac{3}{4}$
53. $\frac{5}{6}$
54. $\frac{1}{2}$
55. $\frac{1}{6}$
56. $\frac{23}{40}$

57. $\frac{17}{28}$ **58.** $\frac{2}{15}$
59. $3\frac{1}{9}$ **60.** $8\frac{31}{35}$
61. $3\frac{9}{16}$ **62.** $\frac{11}{12}$

Page 158

63. $\frac{1}{5}$ **64.** $\frac{1}{4}$
65. $\frac{3}{14}$ **66.** $\frac{2}{15}$
67. 2 **68.** $2\frac{2}{3}$
69. 32 **70.** $6\frac{3}{7}$
71. $3\frac{1}{2}$ **72.** $2\frac{1}{4}$

73. 13 **74.** $7\frac{4}{5}$
75. 3 **76.** $1\frac{1}{3}$
77. $\frac{50}{81}$ **78.** $2\frac{1}{3}$
79. 16 **80.** 42
81. $12\frac{2}{5}$ **82.** $8\frac{1}{2}$
83. $\frac{1}{3}$ **84.** $\frac{1}{25}$
85. $\frac{17}{27}$ **86.** $2\frac{1}{12}$
87. $2\frac{1}{10}$ **88.** 2
89. $\frac{3}{7}$ **90.** $7\frac{11}{13}$

WHAT'S NEXT?

You have just finished *Fractions*, the second book in the Steck-Vaughn series, *Math Matters for Adults*.

The next book, *Decimals and Percents*, provides easy-to-follow steps and practice in working with decimals and percents. The book begins with a Skills Inventory that you can use to discover your strengths and weaknesses, and it ends with a second Skills Inventory that you can use to measure the progress you've made.

In *Decimals and Percents* you will learn to solve real-life problems using decimals and percents. Many everyday measurements are given in decimal form. In working with money, dollars and cents are expressed in decimals. When you buy items on sale, you may get 25 or 50 percent off the original price.

How many other situations can you think of in which you might use decimals and percents? List them on the lines below.

_____

_____

_____